AMATEUR RADIO

PRACTICAL HANDBOOK SERIES

Integrated Circuits: How to Make Them Work
by R. H. Warring

Radio Control for Modellers
by R. H. Warring

Pianos—Care and Restoration
by Eric Smith

Amateur Radio
by Gordon Stokes and Peter Bubb

PRACTICAL HANDBOOK SERIES

AMATEUR RADIO

by
GORDON STOKES
and
PETER BUBB

LUTTERWORTH PRESS
GUILDFORD SURREY ENGLAND

First published 1981
Second edition 1983

ISBN 0 7188 2477 6

Printed and Bound in Great Britain
by Ebenezer Baylis and Son Ltd.,
The Trinity Press, Worcester, and London

CONTENTS

		page
Foreword		7
Introduction. About the Radio Amateurs' Examinations and Licence		9

Chapter

1.	Radio Communication	13
2.	Electrical Fundamentals and Introduction to Symbols used in Radio Circuits	21
3.	Alternating Current — Part I	45
4.	Alternating Current — Part II	59
5.	Semiconductors	68
6.	Transistors — Part I	79
7.	Transistors — Part II	90
8.	Modulation	98
9.	Receivers — Part I	103
10.	Receivers — Part II	116
11.	Transmitters	123
12.	Transceivers	133
13.	Antennas	135
14.	Transmission Lines	145
15.	Propagation	151
16.	Repeaters and Satellites	160
17.	Measurement	162
18.	Safety	173
19.	Operating Procedures and Practices	176
Appendix	List of books published by the R.S.G.B.	185
Index		187

LIST OF PLATES

(between pages 128 and 129)

1a QSL 'Wallpaper'
  b Inside view of QRP (low power) CW Transceiver
2a Rear view of home built equipment
  b Home built SWR bridge and amateur bands receiver
3 Receiver construction
4a Traditional Morse key and Automatic key
  b Dual trace oscilloscope
5 A good antenna tuner
6 & 7 Amateur stations can be complex and expensive
8 Sophisticated SWR and Power meter

FOREWORD

This book aims to help those who wish to study for the Radio Amateurs' Examination, and to explain this fascinating hobby to anyone desirous of knowing a little more about it.

It must be made clear that the book is the work of Peter Bubb, who has been licensed for many years, and is well known and widely respected as a lecturer and coach.

My own function, as a professional writer, has been to edit and to assist where necessary, having obtained my licence quite recently.

I was in fact coached for the examinations by Peter, and as a result managed to pass at the first attempt. If this book can help others to become radio amateurs, we will both be content.

Gordon Stokes,
Bath, March, 1981.

Introduction

ABOUT THE EXAMINATION
AND THE LICENCE

The highly coveted Amateur Radio Licence is issued by the Secretary of State for the Home Office, and must be obtained before a transmitting station can be established.

Licences are issued under the classification 'A' and'B'. The 'A' licence permits the use of all amateur bands and all modes of transmission. The 'B' Licence permits use of all amateur bands above 144MHz, but excludes the use of Morse telegraphy.

Requirements for both licences are as follows: —

1. Be over 14 years of age and of British Nationality.
2. Have submitted the appropriate fee (renewable annually).
3. Have passed both components of the RAE (Radio Amateurs' Examination).

In addition for the 'A' Licence:

4. Have passed the Post Office Morse Test within the previous 12 months.

Radio Amateurs' Examination

This is set by the City and Guilds of London Institute, and is held at centres throughout the country twice each year, in early December and in May. Preparatory courses are available, but these are not mandatory.

The examination is of the multiple choice answer type, so that there is no written work to be undertaken. Candidates are presented with a series of questions, each accompanied by four possible answers of which only one is correct. The correct answer is indicated by the examinee on an answer sheet. Both question paper and answer sheet are given up at the end of the examination.

There are 95 questions in all, two and three-quarter hours being allowed for completion.

The two components of the examination are paper No. 765-1-01, on licensing conditions and transmitter inter-

A*

ference. Paper 765-1-02, on operating procedures and practices, electrical theory, semiconductors, radio receivers, transmitters, propagation and antennas, and measurement.

The first paper contains 35 questions, and the second 60. Candidates are expected to take both components at their first attempt, but success in one may be carried forward, only the unsuccessful one being retaken.

A substantial part of the first paper objective is the demonstration of a satisfactory knowledge of the licence conditions. To this end a complete facsimile of an amateur licence is contained within a very useful booklet entitled *How to become a Radio Amateur*, which is issued free by the Home Office. (Address at the end of this section.)

Information about the Morse test, and the centres at which the test is available, is also given in this booklet.

The Post Office is the examining body in this case, and the speed required is twelve words per minute both in sending and receiving. The length of a word is five letters, and two three-minute periods are allowed for sending and receiving, thirty-six words in each case. Numbers are dealt with separately, ten groups of five figures must be correctly received in a period of one-and-a-half minutes, a further ten groups being sent in a similar period. No punctuation symbols are included in the test.

Since the Morse test is only valid for 12 months, and the RAE passes are valid indefinitely, most candidates will take the RAE first, and then follow with the Morse Test.

The Post Office, apart from supervising this examination, is also the body responsible for monitoring and supervising the whole business of Radio Frequency communication in the UK.

The Post Office also acts as the body responsible for television licence fees and any complaint of interference will be investigated by them. Occasionally, interference to other services will be attributed to an amateur station and under the terms of the licence the Post Office is authorised to inspect and examine an amateur radio station. This is always done in a constructive and helpful manner.

By international agreement applicants for licences must demonstrate a knowledge of basic theory, licence regulations, and their obligation to keep clear of the other, sometimes essential services. For this reason there is more formality and preparation called for than is the case with almost any other pastime.

Anyone taking up Amateur Radio as a hobby should consider joining the Radio Society of Great Britain (RSGB). This society originated in the very early days of radio in 1913, and today represents the interests of some 23,000 members at home and abroad, most of whom are licensed.

Amateur radio is only a very small contender, on the international scene, for the use of the very valuable frequency bands used for communication and broadcasting. The amateur wavebands are jealously guarded. The RSGB is recognised by the Home Office and the various international bodies set up to control world wide communication. It is vital for us individually that we give it our support, so that collectively it may represent our interests at conferences throughout the world where decisions affecting our future in radio are taken. A list of their publications is given in the Appendix.

Full details can be obtained from:

The Radio Society of Great Britain,
Alma House,
Cranbourne Road,
Potters Bar,
Hertfordshire, EN6 3JW.

and the free booklet, *How to Become a Radio Amateur*, from:

Radio Regulatory Dept.,
Radio Regulatory Division Licensing Branch (amateur),
Waterloo Bridge House,
Waterloo Road,
London SE1 8UA.

Chapter One

RADIO COMMUNICATION

Information about all aspects of the world around us is understood by one or more of our senses being activated — smell, touch, sight, sound, or hearing. In amateur radio we are mainly concerned with the sense of sound.

Sound is conveyed by means of waves which activate our ear drums. Let us reconsider this subject, because some important principles are involved.

Sound waves are produced by movement. To produce a sound something must move, and for the sound to travel from source to recipient there must be a medium in which the waves can travel.

In the simple case illustrated in Fig. 1a a tuning fork is struck at point X, which causes it to vibrate backwards and forwards. In so doing it pushes outwards waves of increased air pressure which are followed by waves of decreased pressure. In open space these waves would radiate in all directions, gradually weakening so that the high and low pressures would eventually equalise. The listener hears the sound produced by the fork, by having his ear drum subjected to the alternating pressure waves, the high pressure deflecting the drum to the right, or inwards, and the low pressure waves drawing the drum outwards. In this way the ear drum vibrates backwards and forwards at the same rate as the tuning fork. Providing these vibration speeds lie within certain limits, the deflections are conveyed from drum to brain by a series of complicated bones, tubes and nerves, so that the listener has the sensation of hearing a sound.

A simple experiment has great significance in the present context. Two tins are connected by means of a length of twine which passes through a hole in the base of each, and is retained by a knot. Sounds entering one tin can be heard in the other, provided the string is taut and the tin is placed over the ear.

This apparently trivial experiment reveals important basic

13

communication principles and could be considered as a form of primitive telephone. In this example, one tin would be the transmitter and the other tin the receiver. The taut string is the medium by which the sound information is passed from one to the other.

The 'tin-can telephone' has a very limited range, and the quality of reproduced speech is poor. The range is no further than that produced by shouting in the normal way. In the telephone system the taut string is replaced by lengths of electric wire, which not needing to be kept taut, may wind their way into any corner of the world (Fig. 1b).

There is the story of the simple man who when asked how the telephone system worked said 'It's like a dog, you stamp on his tail, and the bark comes out the other end'. 'What about the radio then?' asked his enquirer, 'Same', said the rustic, 'Cept there ain't no dog.'

The important principles which emerge from this simple example, upon which much communication depends, are:

1. The sound waves were conveyed or transmitted from the fork to the ear by setting up movement in the air, the medium by which they travelled. Sound can also travel in many other media, such as water, fluids, wood, brick, and other solids.

2. The sound from the tuning fork has two components, one is the rate at which the fork prongs are vibrating, and the other is the amount of deflection that is taking place. In Fig. 1a this will be the amount indicated by the dotted lines on the fork. These two components are important because they distinguish one sound from another. The rate of vibration is the frequency, and is measured in Hertz (Hz), being the number of vibrations taking place in one second. The amount of deflection is the amplitude, as amplitude increases so does the level of pressure. This creates increased deflection of the ear drum, and the sensation of sound is louder.

3. The medium that produces sound and conveys it must be elastic, that is it must be capable of being deflected from its original shape and have the ability to return to it. It must also have mass and substance. From this one can see why sound waves cannot travel through a vacuum, and why some materials like flock and felt are sound-deadening, because

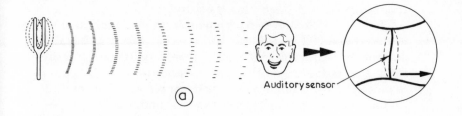

Auditory sensor

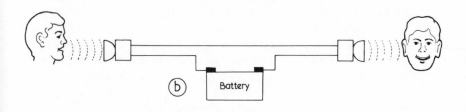

Battery

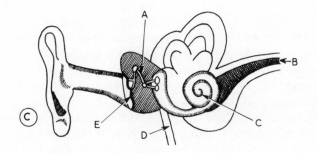

Fig. 1. a Tuning fork producing sound pressure waves
        b. Simple telephone
        c. Simplified view of middle ear
            A. Ossicles transfer vibrations from drum to inner ear
            B. Auditory nerve, carries impulses from cochlea to brain
                hearing centre
            C. Cochlea, coiled tube that converts sound vibrations to
                nerve impulses
            D. Eustachian tube, equalises pressure to outside
                atmosphere
            E. Ear drum membrane, vibrates with sound wave pressure
                variations

they have little mass and poor elasticity.

There is a well known experiment in which an alarm clock is placed in a bell jar with the bell ringing. The base is sealed and the air extracted with a small pump. As the air pressure falls, the sound from the bell reduces until it is almost inaudible. It is difficult under these simple conditions to obtain a complete vacuum so some faint sound will still be conveyed beyond the jar.

4. Not all vibrating masses or frequencies can be heard. The human ear (Fig. 1c) responds to sounds from about 20 Hz to 20 kHz, these limits do vary considerably, however, especially the higher one. Young children hear higher frequencies easily, but this decreases with age. Most of the sound information we receive lies between about 30 and 5000 Hz, and the spoken word can be confined to a much narrower range or band of frequencies than this. This is an important point in radio communication, as will be seen later.

The sounds heard in everyday life are unlike the pure notes which the tuning fork produces. Instead they consist of a complex mixture of frequencies and amplitudes which characterise one sound compared with another.

Any sound can be analysed and reduced to its several components of simple pure tones, but such analysis lies outside the scope of this book. It is sufficient to say that normal sounds are complex mixtures, and the truth of that still holds true in relation to the simple tones.

Turning now to the tuning fork, when struck it vibrates at only one frequency. This is the purpose of the fork, it is used as a reference to adjust other musical instruments to correct pitch. No matter how hard or how lightly it is struck the frequency of vibration is always the same. For middle C on the piano it vibrates at 261 times per second (261 Hz). We have seen that a sound medium must have mass and elasticity, and the amount of these contained in the fork determines its natural frequency, usually termed the resonant frequency.

In many musical instruments the note is produced by a vibrating string being either plucked, bowed, or drawn, for example, the violin. The string is held between the bridge and the pegs. The amount of tension in the string is

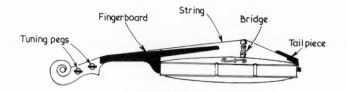

Fig. 2. Side view of violin

regulated by twisting the peg and this determines the degree
of elasticity. The diameter and length of the string and the
material from which it is made determine its mass. Altering
either will cause the natural resonant frequency to change.
When a player strings his instrument he uses thick strings for
the lower notes and thinner ones for the higher notes.
Because of the greater inertia, or resistance to movement, of
those with more mass, they tend to vibrate more slowly. With
the strings fitted, the instrument is tuned to the exact pitch
by turning the pegs, and so altering the tension. By
tightening the string, two things can be made to happen: the
tension increases and with it the elasticity. Also, as more of
the string is wound on the peg, the mass in the remainder
reduces slightly, also contributing to a rise in pitch (Fig. 2).

Finally, to alter the notes rapidly during the course of
playing a melody the player places his finger over the string
and presses it down on the finger board. This is called stop-
ping. What has happened is that the string is now effectively
shorter, it therefore has less mass, and vibrates at a higher
frequency.

If you were able to attach a fine pin to the prong of a
tuning fork and make it strike a line over a moving surface,
the line would follow a particular curve (Fig. 3).

This experiment is therefore theoretical rather than practical since it would be difficult to hold the fork absolutely
steady and the curve drawn would be of small amplitude.
Also, if a pin is attached to the fork its mass will have been
increased and it will vibrate slower or lower than the stated
frequency. However, in the illustration a piece of flat card is
allowed to rotate on a record player turntable, the pin dip-
ped in ink, and the fork set vibrating. The resulting curve

traced would be as in Fig. 3. This type of curve is known as a sine curve or sine wave.

The sine wave will be referred to again during the course of this book, as it is a fundamental part of radio and electronics. The experiment is also interesting in that the reverse of this procedure occurs when we play gramophone records. In this case the grooves are etched into the disc. The pin, or stylus as it would be in this case, rests in the groove, and as the record rotates the stylus is guided rapidly from side to side. The vibrations thus set up in the stylus holder are converted into sound waves by an electronic process which we shall investigate in due course.

In Fig. 3 the points A and B represent the wavelength. That is the distance between two similar points on adjacent vibrations. The actual distances here will depend on how rapidly the fork was vibrating, and the speed at which the cardboard was passing beneath the pin. It follows that if the tuning fork that produced the sine wave in Fig. 3 is replaced by one that vibrates at a lower frequency, and the speed of the rotating disc is constant, the distance AB will increase. This is because the pin is taking longer to vibrate laterally, and in this longer period of time the cardboard is able to travel further.

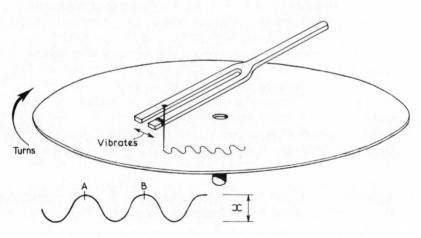

Fig. 3. Wave form produced by tuning fork vibrations

Later on we shall see the importance of this relationship between wavelength and frequency as it is applied to radio.

Sound, by way of speech, or grunts, has been (and some might argue grunting still is) our basic means of communication. It is perfectly satisfactory for intimate conversations, meetings and gatherings. Beyond this we become aware of its limited range. That is to say the distance the unaided human voice can travel and still be understood is small. In radio the word range is used in the sense of distance and the word band to describe a group of frequencies which lie between any given upper and lower limit.

The human voice can travel perhaps a mile under ideal conditions. Its range is limited by the relatively small power that can be imparted to it. The waves of pressure difference gradually equalise, until no pressure differences remain and therefore there is nothing to vibrate the ear drum. Air however is not the only medium which can be used to transmit sound.

Electricity is a medium which can be used to relay sound signals. As we saw earlier, the wire which contains the electricity can pass round corners, through obstacles, and travel great distances. This is how the telephone system operates, and in its simplest form it consists of two transducers connected by wires and supplied with a source of electricity. The transducer is a device for converting sound waves into electric waves or vice-versa.

Note the similarity to the earlier experiment when transducers (tins) were converting the sound waves into vibrations of the string medium. In this example the transducer acts both as transmitter of sounds and as receiver, whereas in practice of course, the telephone handset has one for each purpose.

The next step is to remove the 'dog', or in this case the wires that connected the sender to the receiver. Having done this we arrive at 'wireless' or, radio communication.

Energy itself in certain forms can act as a medium, and can be radiated from a source without the need for any intermediate mass. Heat and light for example radiated from the sun are forms of energy which reach us after travelling millions of miles across space.

Radio waves, which we shall examine in more detail later, are a form of energy which can be radiated in this way. They are capable of travelling great distances at extremely high speeds.

The frequencies at which radio waves vibrate cover a very wide band, and to distinguish them from the audio frequencies we usually refer to them as RF and AF respectively. All RF frequencies are supersonic, that is to say beyond the audible limit. It would be no use therefore to supply them to a transducer in the manner described earlier, since the waves they produce would be inaudible. Audio frequencies are incapable of radiation in the manner of light and heat, only by the medium of air or some other mass as we have seen. Even if it were possible, mass communication would be impractical since every conversation would be taking place in the same band of audio frequency, creating a babel which would be incomprehensible. Imagine all the telephone calls in one exchange going along one wire! Radio communication is never completely free from this problem, and conflicting signals are termed *interference*.

The answer to this dilemma has been the development of radio communication. Thousands of bands of frequencies in the RF spectrum can be isolated and radiated, each separate from the other. These individual small bands of frequencies are sometimes called channels. If now the audio frequencies which we wish to communicate can be superimposed or somehow carried on the radio frequencies, we have solved the problem and created many thousands of usable channels, all capable of being radiated at the same time, each separately distinguishable at the receiving end. This is the carrier principle, the RF signal acting as a vehicle for the AF signal.

The means by which the signals can be formed, transmitted, then separated and received, and their particular application in amateur radio is the basis of the remainder of this book.

Chapter Two

ELECTRICAL FUNDAMENTALS AND INTRODUCTION TO SYMBOLS USED IN RADIO CIRCUITS

It was known in Ancient Greece that when amber was rubbed with silk, it would attract small particles. Other materials will do the same. Try tearing a piece of tissue paper into small sections, then take a plastic rod (a ball point pen will do), rub it briskly on the sleeve of your coat, and it will attract the small pieces of tissue paper to itself from a short distance. Fig. 4a.

The Greek word for amber is ELEKTRON, from which has come our own word of ELECTRICITY or ELECTRONICS. It is intriguing to realise that for centuries this vast source of energy represented by electricity lay dormant, waiting to revolutionise the world.

To understand what happens in the experiment above, and with electricity in general, we must return to first principles.

The smallest part of any chemical capable of individual existence is the atom — the smallest characteristic component of that material. The smallest, lightest and simplest atom is hydrogen, consisting of the central nucleus, a proton, and an orbiting electron.

The atoms of heavier elements contain more protons in the nucleus and have correspondingly more electrons orbiting. Fig. 4d shows a carbon atom, and it will be noted that now there are additional particles in the nucleus; these are neutrons and they occur in all types of atoms with the exception of hydrogen. Their purpose is to bond the protons together, and they are approximately the same in number as the protons.

The electrons and the protons have a mutually attractive pull on each other. This pull or attraction is decribed as an electrical charge. To identify the different nature of the charges the electron is said to be negative and the proton positive. Unlike charges attract and like charges repel. For this reason protons need to be bonded together with

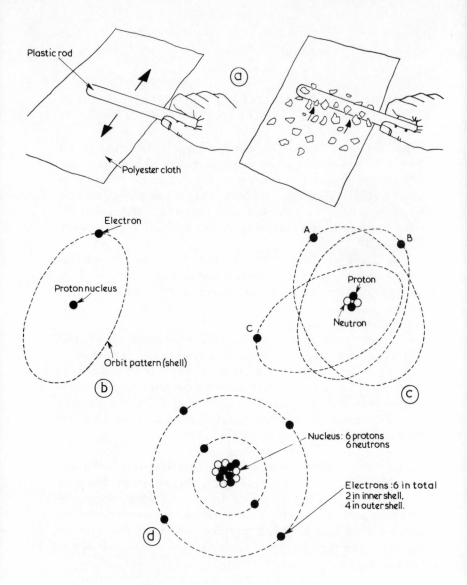

Fig. 4. a. Demonstration of static electricity
b. Hydrogen atom
c. Heavier atoms have more protons and electrons; note the neutrons in the nucleus
d. Carbon atom. Nucleus consists of 6 protons and 6 neutrons 6 electrons orbit, 2 in the inner shell and 4 in the outer

neutrons, which themselves are electrically neutral. In a normal state the protons and electrons are equal in number and opposite in charge so that the overall effect is neutral. However, if for any reason, the atom loses one of its electrons this overall neutrality becomes unbalanced. The nucleus has more positive charged particles than it has electrons to balance them. The atom now has a net positive charge, whilst the detached electron is by itself negative. For this reason our solitary atom and its separated electron mutually attract one another and, with the electron restored to its orbit, the neutral state once more prevails.

The atoms of different elements are composed of the same particles; electrons, neutrons, and protons, regardless of the element they constitute. It is the number of their particles which determines the characteristics and weight of the substance. Different elements vary considerably, however, in the ease with which electrons can escape from the parental orbit. When electrons do escape it is always from the outer orbit or 'shell'. This outer shell is known as the valency band.

One of the elements whose atoms release electrons readily from its valency band is copper. The atom of copper (or any other atom) is extremely small, far too small to be examined by any conventional optical means. In fact many millions of atoms would stretch across the head of a pin. It follows that the electron is infinitely smaller, and in a small piece of copper wire, there are always several millions of 'escaped' electrons drifting in the space which lies between the atoms.

The concept that matter is composed mainly of space is difficult to grasp. Space occurs between the atoms making up the mass, and space within the atom itself. In the piece of copper the drifting electrons move at random and have no overall direction. It is a continuing succession of electrons escaping from an orbit, drifting in inter-atomic space, colliding with other atoms, perhaps dislodging other electrons and taking their places in the orbit; the process producing at any given moment a stock of drifting electrons. A material which exhibits this property is called a conductor, whilst materials which do not release electrons readily because they are tightly held in their orbits are called insulators. Glass is one example. In a later chapter we will look at a very important group of materials which are half way between these two extremes and are called semiconductors.

Examples of Conductor materials:
Copper, silver, aluminium.

Examples of insulator materials:
Glass, mica, ceramic, plastic, air.

Current

In the experiment with the plastic rod, electrons were being knocked out of orbit by the friction of the two materials being rubbed together. In this case electrons from the cloth have accumulated on the rod, thus leaving it with surplus electrons and therefore a negative charge. The rod will then attract the relatively positive pieces of paper, and provided that they are light enough, will be able to hold them to itself.

Static electricity always exhibits itself in insulators because the charge is unable to flow away. In the case of conductors, the charge is able to flow into surrounding materials and become neutralised.

Once the charge moves in this way, it is no longer static and we refer to it as current. Current is the flow of electricity and can be likened in many instances to the flow of water along a pipe. For many people this water analogy renders the fundamentals of electricity easier to grasp. The attracting force created by the charge can be put to use in a conductor.

One method of creating the charge is by rubbing the rod on cloth, but it can also be done chemically, and a device which can produce a charge or potential difference between its terminals is called a battery. If the battery is connected to each end of a piece of copper (which is a conductor), the electrons, whose movements were hitherto random, will be attracted towards the electron deficient—or positive—battery terminal (Fig. 5). The battery here acts like a pump, forcing the electrons collected at its positive terminal towards the negative one, thereby maintaining the pressure or potential difference between the two ends of the conductor. Note one very important fact: the electron current flow in the con-

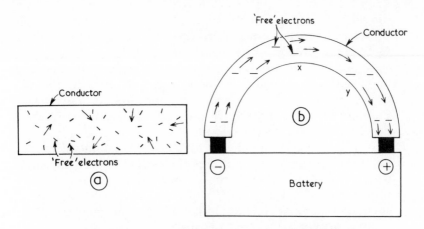

Fig. 5. a. Conductor with no potential difference between ends. Electrons
             drift at random. No overall movement in one direction
         b. Battery. Electrons flow towards the positive terminal

ductor is from negative to positive, thereby causing considerable confusion and misunderstanding. It was an unfortunate
choice of terminology in the early days of electricity because
one naturally tends to think in terms of things moving from
high to low. There is the school of thought which evades this
problem by using the term 'conventional current' flow, which
is the reverse of 'actual current' flow. This can be even more
confusing, since whatever it is called, current flow takes place
in one direction only at any instant. In other words, electrons
move towards the more positive potential between any two
points in a conductor.

In the battery or pump the reverse is happening, the electrons are being forced against their natural tendency from
the positive terminal to the negative one, this is how the battery is able to maintain its charge or potential, and all
generators of electric force are able to do this.

In this situation, the flow of water is from A to B (Fig. 6).
The pressure or potential causing this is gravity, and is
proportional to the height of tank A above tank B. In order
that the flow may continue the water in A must be
replenished by pumping it against its natural tendency,
uphill, thus maintaining the pressure. Most mains water is

distributed in this way with a header tank A providing pressure. In many cases nature takes the place of the pump by filling the header or reservoir with rainfall.

Notice that the pressure is determined by the distance P, and that this gradually reduces along A — B until the B pressure is zero. At the same time the flow of water, the quantity of water moving, is the same at any point in the pipe. So it is with electric current.

In this case the rate of electron flow is the same at every point on the conductor, but here the electric charge pressure is decreasing so that between + and point X it is half the total, between + and point Y a quarter, and there is no pressure difference between two similar points at the + end (Fig. 5).

Some names must now be given to the various factors which constitute the flow of current in an electrical circuit. The units we use and their definitions and derivations belong to the international system of units, or simply S.I. units. For electric current to flow in a circuit there must be a pressure to cause it to flow; this is called an electrical potential and is measured in units known as VOLTS. There must be a conductor to carry the current and, oddly this is not usually measured in units of conductance, but rather of resistance, the unit of resistance being the OHM. In other words, the more ohms a circuit has the less current is likely to flow through it. The current flow itself is measured in AMPERES, usually abbreviated to AMPS.

The water analogy referred to the height of the head of pressure which caused a current to flow. To describe the flow, it is necessary to know how much water is flowing. The quantity used in this case is the litre, and the rate of flow would be so many litres per second.

In an electric current the basic unit of quantity is the COULOMB; this is equivalent to the charge on 6,000,000,000,000,000,000 (six million million million) electrons. The electron is such an infinitesimally small particle, that to obtain a practical sized unit this large number is used. Six million million million is usually written as 6×10^{18} for convenience. The abbreviation of coulomb is Q while for volts it is V and for amps, A. The initial letter is not always used in abbreviations. In order to avoid ambiguity two letters

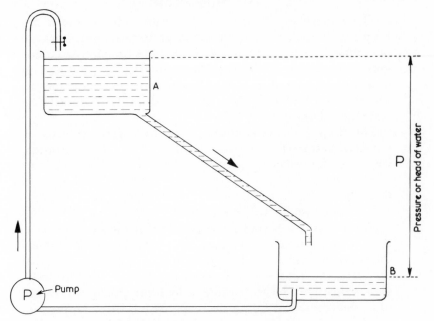

Fig. 6. Water tank analogy. Pump creates pressure in A by forcing water
against gravity — upwards

are sometimes used as in Hz for Hertz, or lower case letters
(as in s for second), or sometimes a letter from the Greek
alphabet. There are several basic quantities contributing to
the make up of electric currents, and it would be very clumsy
to write and read if some form of shorthand were not used.
This is also true of circuit diagrams and symbols.

Electric current flow is a quantity of electrons flowing,
and one coulomb per second is called an ampere. Again, instead of using the rather long-winded 'electric current flow',
the symbol I is employed. So I = current and A = amps; one
is the symbol for current, the other is the symbol for the unit
quantities in which it is measured.

Following on from this the resistance, symbol R, is
measured in ohms, symbol Ω, one of the Greek letters
(omega). The unit of resistance of 1Ω is that offered by a
conductor through which one amp of current is flowing when
the potential difference between the ends of the conductor is
1 volt.

The battery, or device, which creates this difference by supplying the surplus electrons at its negative terminal, is providing what is described as EMF—Electro Motive Force, sometimes simply E. The quantity of force so produced is measured in units of volts. One volt is that electrical potential which will cause one amp of current to flow through a resistance of one ohm.

This important relationship between current, electrical potential and resistance is contained in the first basic formula encountered in electricity, OHM'S LAW.

Ohm's Law

$E = I \times R$, usually written $E = IR$.

This formula is saying algebraically that the electromotive force is proportional to the current. If one increases, so must the other. It also states that the current is inversely proportional to the resistance. If R is increased in the formula above, then in order to multiply it by I and still keep it equal to E, the amount of I must decrease. This is what happens in an electric circuit. If the resistance is increased, and electromotive force remains constant, the current will decrease.

In the above formula, E is measured in volts V, I is measured in amps A, and R is measured in ohms Ω.

Any circuit, simple or complex, where current is flowing, will also have voltage and resistance. There must always be a path of conducting material between the positive and negative potential through which the current can flow.

Resistance. The flow of electrons through a wire is interesting in itself, but of little practical use as such. It is the use made of it, and the work upon which it is employed that makes electricity such a flexible form of power. The first aid to the control of an electric current is the resistor. This component is very common in electric circuits of all descriptions and in a radio or television set probably accounts for half the parts involved. A resistor is a deliberately contrived amount of resistance, usually in the form of a small cylindrical piece of material, colourfully painted and having a piece of conducting wire protruding at each end (Fig. 7).

Many of the resistors encountered in this tubular form are made up of a mixture of carbon paste, which is conductive

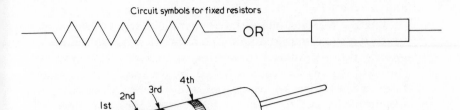

Circuit symbols for fixed resistors

OR

Fig. 7. Typical resistor with coding bands. (Bands are nearer one end and read from left to right). *See* tables attached.

Colour code for resistors.
(Bands are nearer to one end and read from left to right)

1st Band	
Black	0
Brown	1
Red	2
Orange	3
Yellow	4
Green	5
Blue	6
Violet	7
Grey	8
White	9

2nd Band	
Black	0
Brown	1
Red	2
Orange	3
Yellow	4
Green	5
Blue	6
Violet	7
Grey	8
White	9

3rd Band	
Silver	Divide by 100
Gold	,, ,, 10
Black	Multiply by 1
Brown	,, ,, 10
Red	,, ,, 100
Orange	,, ,, 1000
Yellow	,, ,, 10,000
Green	,, ,, 100,000
Blue	,, ,, 1,000,000

4th Band Tolerance	
Red	± 2%
Gold	± 5%
Silver	± 10%
No Colour	± 20%

and a type of clay paste which is non-conductive. By controlling the relevant quantities of each, a mixture is obtained which has the required amount of resistance. When moulded into the little cylinders a connecting wire is embedded in each end, and the resistor is then baked. Some resistors are made from special wire known as resistance wire, which has a specific resistance for a given unit length, and therefore a very accurate resistor can be produced by cutting the appropriate length of wire and winding it into a coil, usually on a base of some insulating material. There are also carbon film resistors, made by depositing a layer of carbon of varying thickness, and hence resistance, on to a ceramic tube.

The value of the resistor in ohms is sometimes painted on to the body of the component together with other information, including its accuracy, expressed as a percentage. More often the value is encoded in the form of coloured bands (Fig. 7). Resistors also exist in different physical sizes. This is because when passing a current a resistor will become warm. In certain applications it may become very hot, and the fabric of the component will deteriorate or even disintegrate. In either event, the resistor no longer has its original value. Therefore, they will be selected in a size capable of dissipating the anticipated heat without any change in the value. Wire wound resistors are generally capable of withstanding much greater heat than the carbon type. They cannot however be used in some applications where they would otherwise be very useful, because of the way they behave in the presence of certain forms of current, as will be explained at the appropriate point.

Lastly, there is the variable form of resistor, known as a potential divider, or potentiometer ('pot' for short), so called because it can divide the potential voltage between its two ends into any two values, the sum of which is equal to the total, by varying the position of the sliding contact (Fig. 8). Again, these can be wire wound or carbon track, but mostly they are carbon. It is a common component, and the one we most often contact with outside the equipment we use. Potentiometers are used to control volume, bass, treble, brightness, colour and so on, in Hi-Fi sets and televisions. The device itself is inside the cabinet, a spindle being extended and a

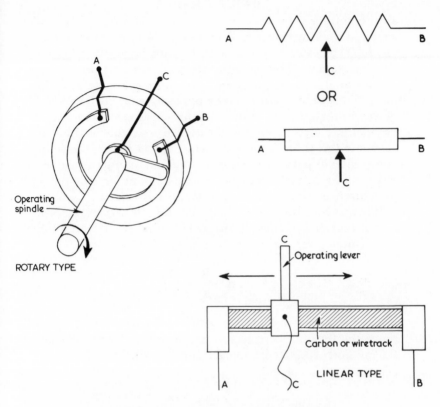

Fig. 8. Variable potentiometer with circuit symbols

knob attached, so that sliding or rotating can be effected.

Any electric circuit has resistance and continuity. It is practical however to assume when considering circuits that the conductors connecting the components have no resistance in themselves. This makes it easier to read a circuit and to make calculations. Where two or more resistors are connected in line, so that the current path is from one to the next they are said to be connected in SERIES. Where resistors are connected side by side, so that the current can divide, some flowing through each path, the resistors are connected in PARALLEL. They can of course be connected in a combination of series and parallel. In the series example, Fig 9a, it is easy to calculate the total resistance in the circuit because

31

all the current must flow through each of them and be subjected to their opposition. Therefore to find the total, the individual values of R are added, the sum being the total circuit resistance. The parallel arrangement is rather more complicated because the current has more than one available path. Thus each resistor is only opposing that current which flows through it, so that the more resistors there are in parallel in a circuit, the less the overall resistance will be. To calculate this value is more difficult. If two resistors have the same value in parallel it is easy—an equal amount of current flows through each, and twice the amount of conducting path has been provided, so the resistance is halved.

When two resistors are in parallel (Fig. 9b). then the combined resistance of them is always less than either, and a simple formula gives the result:—

$$\text{R total} = \frac{\text{R1} \times \text{R2}}{\text{R1} + \text{R2}}$$

Taking this example where both are the same value we can prove the formula by calculating the result. Assume R1 and R2 are both 100. Replacing R1 and R2 with these values we get:—

$$\frac{100 \times 100}{100 + 100} = \frac{10,000}{200} = \frac{100}{2} = 50\Omega$$

One more example, Fig. 9c, this time to show how the formula can be used to deal with three (or more) resistors in parallel.

Power. When a current flows through a resistor the energy it uses in overcoming the resistance is given up mainly in the form of heat, but some is used in creating a magnetic field. Sometimes the heat is the desired effect from the current flow, as in the bar of an electric fire or the filament of a light bulb, which is simply a piece of wire at white heat. More often in electronic circuits the heat is an undesirable by-product, rather than a purposeful function of the current. Measures must be taken either to convey this unwanted heat away,

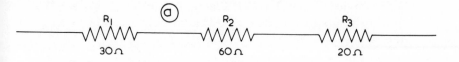

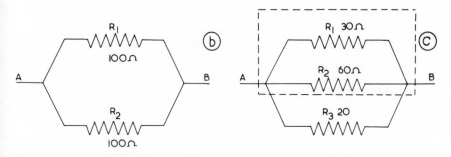

Fig. 9. a. Resistors in series total value 110Ω
       b. Resistors in parallel
       c. More than two parallel resistors. To calculate breakdown into
          boxes of pairs. Total resistance within dotted box

$$\frac{30 \times 60}{30 + 60} = \frac{1800}{90} = 20$$

Remaining resistance calculated thus:

$$\frac{20 \times 20}{20 + 20} = 10Ω$$

or to ensure that the components in a circuit are able to function properly at the temperatures they attain.

Heat is usually wasted energy in electronics; once created it cannot be easily recovered in some other useful form. This brings us to the question of POWER. Wherever energy is converted from one form to another work is done and POWER expended. Power is the name given to the rate at which work is being done. For example, a weight is lifted from the floor to the table top slowly, and a second similar weight quickly. The work done in both cases is the same but the rate at which the work is done in the second case is greater, and therefore more power is expended. In an electric current, the flow of 6×10^{18} electrons per second is one amp, and the rate of flow is caused by the potential difference in volts. So power is volt amps. In fact there is a name for the unit of power, too: 1 volt × 1 amp = 1 WATT, symbol W, and the symbol for power = P. The formula can now be written: —

$P = E \times I$ where P is in watts, E in volts and I in amps.

There is no reference to resistance in this equation, but as we have said, any circuit has resistance as well as current and voltage. From Ohm's Law we saw that $E = IR$, so this can be substituted in the formula as follows: —

$P = (IR) \times I$, or $P = I \times R \times I$, $I \times I = I^2$, so finally

it becomes: —

$$P = I^2R$$

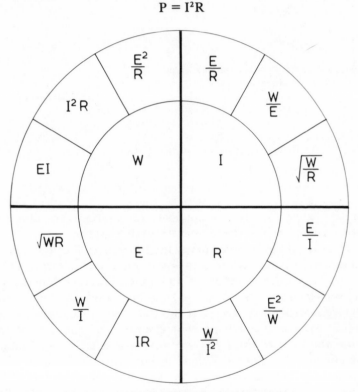

Fig. 10. Formulae deriving from OHM'S LAW
W = Power measured in Watts
I = Current measured in amps
E = Electromotive force measured in volts
R = Resistance measured in Ohms

There are a number of ways in which Ohm's Law and power can be transposed. They can all be put into a simple chart where each quantity in the centre section can be found by the equations in the outer section (Fig. 10.).

As with engines and motors, whose unit is horse power, power is often used to describe the potential available in electrical matters.

We talk for example of a hundred horse power motor even when it is stationary and developing no power at all. So it is with a hundred watt light bulb, or a five watt resistor.

Magnetism

Magnetism is a force which is very difficult to describe. For centuries man has been aware that a substance called lodestone would attract small pieces of iron to itself. The earth is a natural magnet with a north and a south pole, and lodestone is in fact iron oxide which has become magnetised by the earth. Most of us are familiar with bar magnets, in one position their unlike poles will attract one another, whilst the like poles will repel. This is the clue to magnetism and its force. Magnetic fields and electric current are inseparable, one is part of the other.

The area of force surrounding a magnet is known as a magnetic field. It is a space charged with energy, and each field has a North and South pole.

Only a few materials exhibit the property of magnetism, and these are mainly of the ferro-magnetic group. Iron is probably the most common. A well-known experiment demonstrating the area and density of a magnetic field is illustrated in Fig. 11. The bar magnet is placed under the paper, which is not itself attracted by magnetism, but allows the force to pass through it and attract iron filings which are sprinkled on the surface. The paper is tapped gently, whereupon the filings move into the pattern shown in the picture. The density of the filings nearest to the magnet at the pole ends indicates where the field is strongest. The magnetic field is described as having lines of force, and these can be seen spreading out from pole to pole, weakening as they do so. Where the lines of force are intense, the field is strongest.

Electromagnet. If a piece of conductive wire is passed

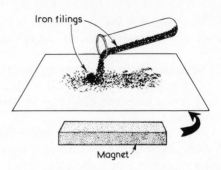

Iron filings

Magnet

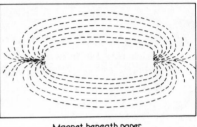

Fig. 11. Experiment to demonstrate the presence of magnetic lines of force. Note the concentration of filings around the pole areas

Magnet beneath paper

through a card and a current passed through the wire, a magnetic field is created round the wire. Again, the filings will indicate that the field is strongest near the surface of the conductor, weakening as it gets further away. This relationship between magnetism and an electric current is most important in electronics. Note also that the conductor creating the magnetic field need not itself be one of the ferrous materials. In fact it is most likely to be copper which is unaffected by magnetism. When the current is switched off, the field collapses.

The field produced in this way is relatively weak unless large currents are used. One way of increasing the effectiveness of such a magnet is to wind the wire into a coil. In this way all fields lie alongside each other so that the effect is enhanced, and quite strong magnets can be produced. The strength of the field is improved further if the conductor is wound on to a core of soft iron. As with all magnets, there is a north and south pole (Fig. 12). This form of electromagnet is called a SOLENOID. Some of the many uses to which this force can be applied will be explained later in the book.

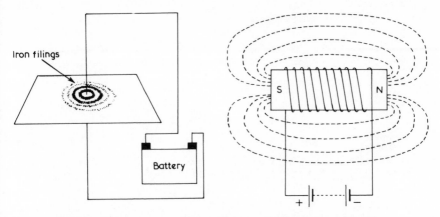

Fig. 12. Electromagnet. If wound into a coil it becomes a solenoid

Soft iron does not retain its magnetism. Steel, however, does, and so permanent magnets are made by exposing steel bars to the field of an electromagnet, whose force they then retain.

Inductance. Current produces a magnetic field. Faraday, in 1831, discovered that a magnetic field can be made to produce, or induce, a current in a conductor. His experiment showed that a *changing* magnetic field would induce a changing current to flow. This is the important difference. The steady current produces a steady magnetic field. But providing a steady magnetic field in the presence of the conductor will not create a current flow.

In the first experiment, Fig. 13, the magnet is dropped from above, so that it passes down through the coil and out at the bottom. In series with the wire of the coil, and making a complete circuit, is an ammeter. This little device has a pointer which moves, indicating the presence of a current. It is in the centre of the scale at zero current and will move to either side, depending on the direction of the current. When the magnet is released, the magnetic field to which the coil is exposed increases in intensity. Therefore it is changing. The conditions for induction are satisfied, and a current flows in the coil. The current too will be changing, increasing as the magnet approaches. There is a moment when it is in the centre of the coil and the field is stationary, where it is becoming

37

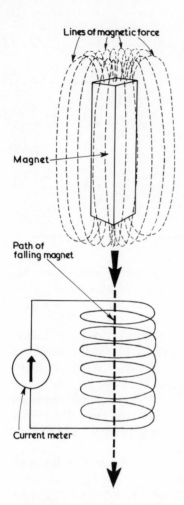

Lines of magnetic force

Magnet

Path of
falling magnet

Current meter

Fig. 13. Induction. As the
magnet falls a current
is induced in the
coil

neither stronger nor weaker. At this point there is no current
flowing. Then the magnet moves on, the field is changing by
weakening, and current flows in the opposite direction,
decreasing to zero as the field moves beyond the coil.

In the second experiment (Fig. 14), when the switch is
opened or closed a momentary flicker of current is indicated
in the second, or secondary, coil. This is because although
the current is usually steady, it takes a fraction of a second to
build up from zero to its nominal value, so during this time

38

the magnetic field it creates in the first coil is also building up. Again, conditions are right for induction. The second coil within the influence of the field of the first, enhanced by the common iron core, and a current is induced in it. When the switch is opened a current in the opposite direction is induced in the secondary winding.

Whenever a conductor finds itself within the changing lines of force of a magnetic field a current will be produced. This applies equally to the conductor producing the magnetic field. In the second Faraday experiment, Fig. 14, the primary coil is creating a magnetic field as the current builds up when the switch is closed, and so conditions exist for a current to flow in itself, due to this effect. This is known as self in-ductance, or simply inductance. Self induced currents are in the opposite direction to the primary currents and of course weaker and, as the current cannot flow in the opposite direc-tion in the same conductor at the same time, it has an op-posing effect. The result is a smaller net primary current to begin with. This is known as the back EMF.

Capacitance. If someone asked you what capacity your ket-tle has, you would understand that they wanted to know how much water it would hold. The average kettle holds about three pints or so. Capacity in electrical terms is the amount of charge a body or conductor will hold. Any conducting wire

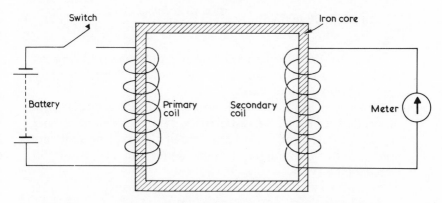

Fig. 14. Faraday's experiment. The iron core is common to both coils and therefore magnetic lines of force extend from primary to secondary coil

therefore has capacitance, and indeed inductance, but in both cases the amount is very small and usually ignored in simple circuits. Just as we set out to make deliberate use of the magnetic effect by winding the conductor into a coil, it is with the deliberate creation of larger amounts of capacitance that we are concerned with now.

Quantities of electricity (charged electrons), are measured in coulombs, and it is the quantity of charge electrons that a capacitor can store, proportional to the voltage supplied, that determines its capacitance value. The symbol for capacitance is C. The units we use to measure it in are farads, symbol F.

This unit, named after Michael Faraday, is another of those unfortunate accidents of history, which in this case has given us a unit far too large for practical purposes. The actual values of capacitance occurring in radio and electronics are thousands or even millions of times smaller than one farad, so for convenience the sub divisions which are used are as follows: —

$$\frac{1}{1,000,000} F = \mu F \text{ a micro farad}$$

$$\frac{1}{1,000,000,000} F = nF \text{ a nano farad}$$

$$\frac{1}{1,000,000,000,000} F = pF \text{ a pico farad}$$

This is rather like the basic unit of fluid measure being the cubic mile, so you might order from your milkman two pico cubes of milk each morning.

Capacitors are made in a number of different ways, but basically they consist of two plates separated from each other by an insulator (Fig. 15). The circuit diagram symbols are also given. The factors which determine the value of capacitance are : —

1. The area of the plates that store the charge.
2. The distance between the plates.
3. The material of which the insulation between them is made.

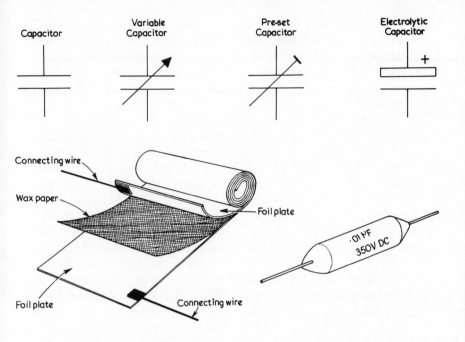

Fig. 15. Paper capacitor construction and the various capacitor circuit symbols

Fig. 15 shows that the capacitor represents an open circuit between its two ends, and if there is a potential difference between them no continuous current can flow, so no power is dissipated in the form of heat, and therefore no power factor need be considered as with resistors. However, a charge will develop across the two plates, the amount of which will depend on their area and the distance by which they are separated, together with the potential difference. If the voltage is increased enough the point is reached where it will jump across the gap. Any insulator will conduct, given sufficient electrical pressure. Thus there is a voltage limit to any given capacitor which must not be exceeded if the component is not to fail. This information, together with the value, is either printed on the body of the capacitor or colour coded.

B*

41

In the smaller value capacitors, the insulating material is often mica or ceramic, with the two plates either side and the entire sandwich immersed in wax or resin to keep out damp. Any deterioration of the insulating material may result in leakage of current between the plates, and loss of capacitance.

Larger values of capacitor are made by having the plates of flexible aluminium in strips with waxed paper between which are rolled up into a tube. Very large values up to 5,000 μ F or more are too bulky when made in this manner, and a type of construction is used which is known as the electrolytic capacitor. The principle of this type rests on substantial reduction of the thickness of insulating material in the sandwich, thus creating more capacitance for a given area. In manufacture the plates are immersed in a conductive fluid compound. When a voltage is applied across the terminals a chemical action is set up which forms a very thin dielectric or insulator on one of the foils. This is thousands of times thinner than paper, and large values of capacitors can be formed in a relatively small space. The process is known as elec-

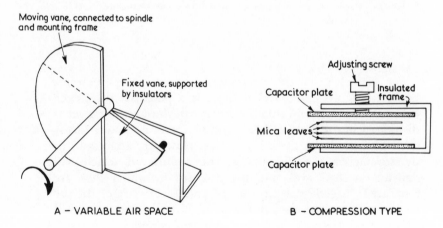

A – VARIABLE AIR SPACE   B – COMPRESSION TYPE

Fig. 16. Variable capacitor
   a. is air spaced and varies in proportion to the area of overlapping plates
   b. is interleaved with mica and varies in proportion to the distance between the plates

trolytic action, hence the name electrolytic capacitor. These capacitors are polarised when they are formed and this polarity must not be reversed. They are always clearly marked positive and negative, and very often the positive terminal is coloured red. An inherent characteristic of electrolytics is that they have a small leakage current, but this disadvantage is offset by the larger values that can be constructed within a manageable size.

It must be emphasised that large capacitors can hold a charge for some considerable time, several days if in good condition, and should be treated with respect. Large value electrolytic capacitors, in spite of their leakage current, can give a nasty shock some time after the equipment they form part of has been turned off and should be discharged through a resistor before handling.

If a capacitor is connected to a battery as in Fig. 17 the meter will indicate any current flow. Assuming the capacitor to be uncharged at the moment the switch is closed (A), a current will flow, its presence and direction being indicated by the meter. The capacitor will quickly charge up so that the voltage across its plates is the same as that of the battery. The time required for this to occur will increase with the increase in size of the capacitor. As it reaches the fully charged condition the current flow ceases, and the meter needle returns to the centre. A good quality component will remain in this condition for some time. The plate connected to the negative terminal of the battery has become full of charged electrons, and a force or potential difference exists across the

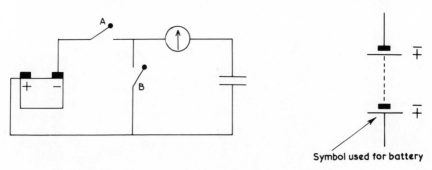

Fig. 17. Circuit to charge and discharge a capacitor

capacitor. If switch A is opened, and B closed, a path is provided for them to flow to the other plate and equalise the pressure. Whilst this current flows it will be indicated by the meter, but now flowing in the opposite direction to when it was charging.

In a circuit such as this, there is normally no current flow after the initial charging period, the capacitor is a break in the circuit to this type of current which is known as DC— DIRECT CURRENT, or undirectional current, flowing or tending to flow in the direction determined by the battery polarity.

The charge stored in such a capacitor is proportional to the size of the capacitor and the potential difference across the plates. The formula is:

$$Q = CV \qquad \text{where } Q = \text{charge in Coulombs}$$
$$C = \text{capacitance in Farads}$$
$$V = \text{PD in volts.}$$

This formula can be rewritten, or transposed in terms of either C or V.

$$C = \frac{Q}{V} \qquad \text{or} \qquad V = \frac{Q}{C}$$

The word 'condenser', meaning capacitor, is sometimes used, but this is an old term not often encountered today.

Chapter Three

ALTERNATING CURRENT — Part I

The current so far considered has been DC which if represented on a graph would appear as a straight line. The amplitude of the line indicates the current. The horizontal scale shows the period of time during which it is flowing. When the current is switched off, the current line descends vertically to zero. If the switch were continually switched on and off, say every half second, the resulting graph would appear as (in Fig. 18) a series of current periods of half a second, alternating with periods of no current for half a second. The graph is useful because it gives three pieces of information about the current.

1. The amplitude of the current.
2. The length of time during which it flows in each period, and the rate at which the current is switching on and off.
3. The manner in which the current is changing.

The shape of the graph (this one is known as a square wave form for obvious reasons) tells us that the current appears instantly on switch-on and disappears instantly when switched off. This type of graph can be used to illustrate another form of current. AC (ALTERNATING CURRENT) is a current that varies in amplitude sinusoidally (Fig. 19a). The term sine wave is applied to these types of curve because they develop from a mathematical relationship of an angle known as its sine. In simple terms this means that the quantity depicted by the graph is changing evenly first in one direction and then in the other. Such a change would in fact be produced by a rotating current generator.

A simplified form of generator is shown in Fig. 19b. The loop of wire is rotating within the magnetic field between the ends of the magnet. For most of the rotation, the wire forming the loop is passing through a changing magnetic field, and so the conditions exist for a current to be induced in the

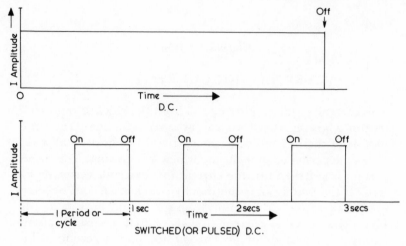

Fig. 18.

conductor. The current thus induced will change direction twice in every revolution, alternately flowing clockwise then anti-clockwise in the loop. The reason for this is that the magnetic field through which the loops pass is first increasing and then decreasing in intensity.

In order to produce more current more efficiently than in our simple loop, in practice, a coil of many turns wound on to a form is used, and allowed to rotate within the field. The shape of the magnet is such that the coil rotates as closely as possible to the surface where the lines of force are strongest (Fig. 19c). The lines of force to which we refer so often in magnetism are known as the FLUX.

Alternating current occurs throughout radio and transmitter topics and will basically always be considered in the form of sine waves. There is, for example, the low frequency of 50Hz, at which our domestic electricity is distributed. (Hertz = cycles per second). Then there is the band of audio frequencies that we considered earlier extending up to about 20,000Hz (20kHz), and sounds can be considered to be a mixture of sinusoidal wave forms. Lastly there are the radio frequencies which extend to beyond 30,000 million Hz — 30,000 MHz.

At this point it may be as well to consider some of the

46

various abbreviations used for very large and small quantities, and in the field of radio it is frequently necessary to deal with either very large numbers or very small fractions.

A thousand units	10^3	is prefixed	kilo	symbol k
Amillion units	10^6	,,	mega	,, M
A thousand million	10^9	,,	giga	,, G
A tenth of a unit	$\frac{1}{10}$	,,	deci	,, d
A thousandth	$\frac{1}{1,000}$	,,	milli	,, m
A millionth	$\frac{1}{10^6}$	,,	micro	,, μ
A thousand millionth	$\frac{1}{10^9}$	,,	nano	,, n
A million millionth	$\frac{1}{10^{12}}$	,,	pico	,, p

To avoid writing down many noughts when these large figures are encountered, the device of the power of ten is used. This simply means that the figure ten is written and the indice, the small figure above it and to the right, indicates how many noughts exist in the number. Thus, 1,000 can be written as 10^3, 6,000 can be written as 6×10^3. Fractions, i.e. $\frac{1}{1000}$ can be written as $\frac{1}{10^3}$, or $\frac{6}{1000}$ can be written as 6×10^{-3}. Note the negative indice in this last configuration which is the most popular way of writing fractions in this form.

Referring again to Fig. 19a the frequency of an alternating current will be the number of complete cycles occurring every second. The amplitude at peak value is that instantaneous value achieved at A. If AC is passed through a circuit containing a lamp, the lamp will light, thereby indicating that power can be dissipated by AC as well as DC.

We need now to show what value of current to use when working out formulae containing alternating currents and voltages. It is obvious that the peak value is not true since it only occurs momentarily every cycle. Some sort of average between zero and peak must be used. It has been found that the root mean square (RMS), that is $0 \cdot 707 \times$ peak value, gives the

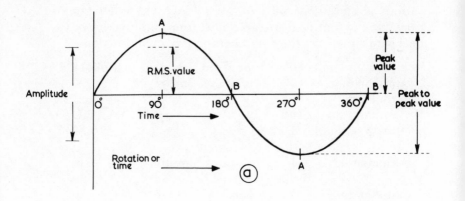

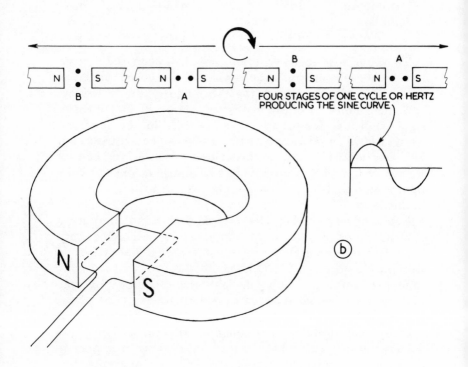

FOUR STAGES OF ONE CYCLE OR HERTZ PRODUCING THE SINE CURVE

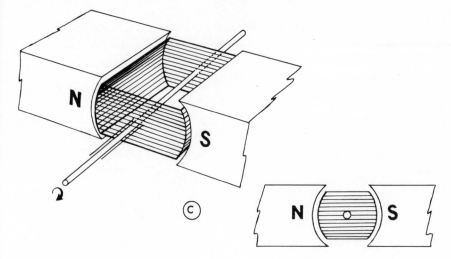

Fig. 19.   a.   AC sine wave
          b.   Simplified generator
          c.   Generator with coil of many turns will produce more current

same power as a DC of that amplitude and is the one used to define AC.

Two or more currents can exist at the same time in supplying a circuit. If two generators are coupled together, so that their output is in step, or in phase, the two sine waves can be added together to produce a sine wave of greater amplitude. Fig. 20a.

If, however, they are coupled together in anti-phase, exactly 180° or a half turn apart from each other, and the output from each is equal, the result will be zero since they are cancelling each other out. (Fig 20b).

The two generators could be coupled at any angle between these two extremes, the difference between them is called the phase angle. If they are 90° out of phase, as in Fig. 20c, the result is a mixture of part addition and part subtraction, at different phases of the cycle. It will be noted that the peaks of the resultant are not equal to the two sine waves added. This can only happen when they are in phase. It occurs half way between them at 45° phase difference. Two alternating

49

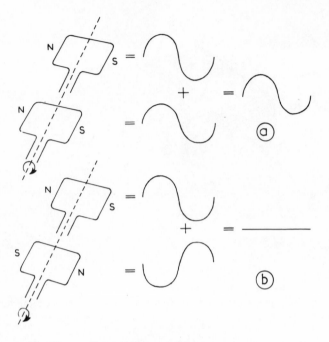

Fig. 20.　a.　Two generators in phase
　　　　　b.　Two generators in antiphase
　　　　　c.　Two generators 90° out of phase

currents out of phase can be indicated more readily by means
of a Vector Diagram (Fig. 21). Here the length of the line OA
is proportional to the current from one generator, and on the
same scale the line OB is proportional to the current from the
other. They are said to be rotating anti-clockwise and the
angle of the two lines to each other is the angle of phase between the two generators. The length of the line OR gives us
the resultant amplitude, and phase angle of the resultant
current. Vector diagrams are more simple to examine than
the more complicated mixture of sine waves that they
represent.

For those who remember Pythagoras, the length of OR is
$\sqrt{OB^2 + OA^2}$, the square root of OB squared plus OA
squared.

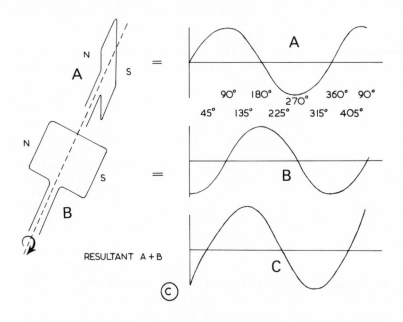

If, for example, OA = 10 amps, and OB = 10 amps, then the resultant will be: —

$$\begin{aligned}
&= \sqrt{10^2 + 10^2} \\
&= \sqrt{100 + 100} \\
&= \sqrt{200} \\
&= 14 \cdot 14 \text{ amps.}
\end{aligned}$$

It is what happens when capacitance or inductance are present in an AC circuit that makes the circuit behave in a manner fundamentally important in radio theory.

In the DC circuit containing a capacitor, there is only a small period of time after switching on (while the plates charge up), during which current flows. Thereafter it behaves as an open circuit allowing no current to flow. It behaves in other words like an infinitely high resistance.

However, with a current alternating in one direction and then the other, the capacitor can continue to charge and discharge, reversing its polarity in step with the alternating current which supplies it. The amount of current flowing will depend on two factors, the size of the capacitor (the largest it

51

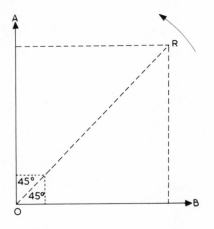

Fig. 21. Vector diagram
O-B = Current from
generator B
O-A = Current from
generator A
the length of O-R gives
the resultant amplitude
and phase angle

is the more charging current it will accept), and the frequency at which the charging current is changing direction. In the DC circuit containing inductance there is a small period of time after switching on during which the current reaches its maximum level, thereafter it behaves as a low resistance allowing a large steady current to flow. The reason for the small delay in the current rising initially is the effect known as the back EMF of the coil. As we have seen, when a conductor is in the presence of a changing magnetic field, a current will be induced in it. The self induced current which arises when the DC circuit is first switched on and the current is rising, causing a changing magnetic field, is in the opposite direction, opposing the primary current. This opposition has to be overcome, and is the cause of the delay.

In the circuit with an alternating current, the current does not reach a steady level, it is continually changing and so, therefore, is the magnetic field around the coil. This means that there is always a back EMF changing in polarity as the current changes, and always opposing the flow of primary current. The amount of current which is able to flow will depend on the frequency, and the number of turns of the coil, since both will affect the rate and the amount of changing magnetic field. The faster the current changes and the greater the number of turns on the coil, the more opposition is offered. The name we give to this opposition is REACTANCE. It cannot be called resistance because resistance is the term for opposition to current flow and is constant, regardless of whether it is DC or AC of any frequency.

The symbol for reactance is X. We use Xc for the capacitive reactance and XL for the inductive reactance. The units of reactance are ohms (Ω) once again. Do not confuse reactance with resistance even though the same units are used for both.

Definition of Capacitance and Inductance

One farad of capacitance is said to exist between the plates of a capacitor when, charging it with one coulomb of electricity, causes a potential of one volt to appear across the plates.

As noted earlier, practical units of capacitance in radio circuits are very small fractions of a farad, and so we use the quantities micro, nano and pico farad. One henry of inductance is said to exist in a conductor when the current in a circuit is varying at the rate of one amp and produces an EMF of one volt.

In radio work, the values of inductance likely to be used vary from several henries down to micro henries.

When capacitance is present in a circuit of alternating current the voltage maxima, or peaks, occur at different times from the current peaks. At the commencement, there is no charge across the capacitor, and therefore the voltage across the plates is zero. At this point maximum current will flow into the capacitor, gradually diminishing as the point of full charge is reached. By this time the voltage across the plates has reached its maximum, so maximum voltage occurs later in the cycle than maximum current. In fact, it is 90^{0} later, or lagging the current maximum. It continues to lag 90^{0} out of phase as the alternating cycle continues and the capacitor charges with the opposite polarity (Fig. 22).

We saw earlier how the size of the capacitor determines how much charge it will accept and thus the current flowing into it. Also the frequency at which the plates are charged and discharged. Both of these factors determine the amount of reactance the component has.

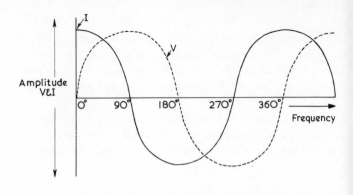

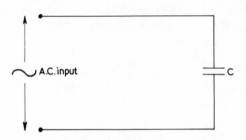

Fig. 22. Circuit with capacitance and alternating current showing current and voltage phase difference

The formula for this is: —

$$X_c = \frac{1}{2\pi fc}$$

This formula shows that the larger either frequency or capacity is, the smaller the quantity becomes. Thus capacitive reactance is inversely proportional to frequency and capacitance.

When an inductance (L) is present in a circuit of alternating current, the voltage and current waveforms are still 90° out of phase with each other, but in this case the voltage occurs first and therefore leads the current. In Fig. 23 at the instant of switching on the full voltage appears across the coil L because of its reactance due to the back EMF. As this is overcome and the current increases, the reactance diminishes and the voltage falls. The current maximum occurring 90°

54

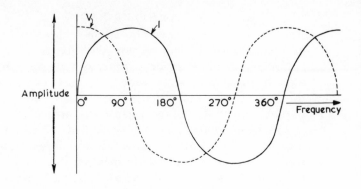

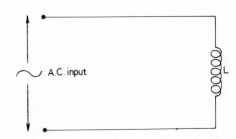

Fig. 23. Circuit with inductance and alternating current showing current
and voltage phase difference

after the voltage maximum. The alternating cycle continues,
the polarity of the current changes and the same happens in
the opposite half cycle.

The formula for the reactance of an inductor is: —

$$X_L = 2\pi f_L$$

Notice again that the frequency as well as the value of the
component are present in the formula, and in this case the
larger either quantity is, the larger the reactance becomes.

Inductances in series or in parallel configurations can be
treated in the same way as resistors — provided there is no
mutual coupling! That is to say if they are in series the total
reactance will be the sum of the individual reactances. In
parallel we can use the form: —

$$\text{Total reactance} = \frac{XL1 \times XL2}{XL1 + XL2}$$

55

treating more complex combinations by breaking them down into pairs as we did with resistor combinations.

Capacitors in series or parallel are treated the opposite way. In parallel they are added, the total capacitance being the sum of the individual values. With two capacitors in parallel this makes sense because the area of the plates has simply been enlarged by the amount of the two areas added together. Capacitors in series on the other hand become smaller because the dielectric between the two end plates of the series becomes greater. So for capacitors in series the formula is: —

$$\text{Total Xc} = \frac{\text{Xc1} \times \text{Xc2}}{\text{Xc1} + \text{Xc2}}$$

In a circuit where inductance and capacitance are in series together with a resistance Fig. 24 we have reactance Xc and reactance Xl plus the resistance of R. In such a circuit, containing both reactance and resistance, a new word is used to describe the opposition to current flow, and to distinguish it from the other two words: this is IMPEDANCE. The symbol is Z and again the units are ohms. (Ω).

In Fig. 24 the voltage is 90^0 leading the current across the inductor and is 90^0 lagging the current across the capacitor. In such a series circuit the current can only be going in one direction at any instant and so voltages across the capacitor and inductor are 180^0 out of phase with each other. If at a particular frequency, known as the resonant frequency, the two reactances are equal, then the voltage will be equal too. Being opposite in polarity, however, they will cancel each other out. This being the case the resultant reactance at this point will be zero, leaving the resistance R as the only opposition to current flow. From Ohm's Law: —

$$\text{Impedance X} = \frac{E}{I} = \frac{O}{I} = O$$

Here is a very important feature of the series circuit, that when the reactances are equal it is said to be resonant, and at resonance it has a very low impedance. At this frequency it will also allow a large current to flow compared to that which will flow at other frequencies. Because of this, a series circuit

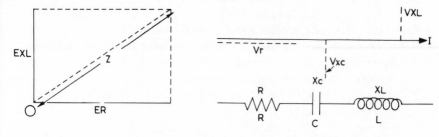

Fig. 24. Series circuit with inductance capacitance and resistance all
present. Vector diagram shows resultant impedance Z

tuned to resonance is sometimes referred to as an acceptor
circuit.

There is only one frequency at which given values of inductance and capacitance will resonate. Since, as the
frequency changes, the reactance of the two components also
changes; one becomes higher, the other lower. Inductive
reactance increases as frequency increases, and capacitve
reactance decreases as frequency increases.

The formula for finding the frequency at which capacitance
and inductance will resonate is: —

$$f = \frac{1}{2\pi \sqrt{LC}} \text{ Hertz.}$$

At any other frequency the reactances are not equal, but as
the voltages across them are 180° out of phase, the resultant
will be the difference between the two. This leaves a voltage
across the reactance which is 90° out of phase with the
voltage across the resistance. This resultant can be obtained
from a vector diagram. In Fig. 24 the difference between the
capacitive and inductance reactances can be considered as
leaving a net inductive reactance. The horizontal line
represents the voltage across the resistance R.

The vertical line represents the voltage across the net inductive reactance. The resultant impedance Z will be the
length of the diagonal line across the completed
parallelogram. Using the formula of Pythagoras again, we
get:

$$Z = \sqrt{X^2 + R^2}$$

57

X will be the balance of the capacative or inductive reactance when one is subtracted from the other.

One way of remembering the phase lag or lead relationship in inductive and capacitive circuits is to use a mnemonic 'CIVIL'. This reminds us that in a capacitive circuit I (current) leads V (volts), whereas in an L (inductive) circuit V (voltage) leads I (current).

Chapter Four

ALTERNATING CURRENT — Part II

In the circuit shown in Fig. 25 the inductor and the capacitor are in parallel and for this reason it is known as a parallel tuned circuit. To calculate the impedance of such a circuit we examine the vector diagram again, assuming for the moment, no resistance. In this case we have to take the instantaneous voltage as being the component with which the phases can be compared. Since the voltage between A and B obviously appears across L and C at the same instant, the current in the capacitive component is in anti-phase with the current in the inductive component, and the net current will therefore be the difference between the two (Fig. 25). The impedance is then found in the usual way from Ohm's Law:

$$Z = \frac{E}{I}.$$

When the two reactances are equal, the circuit is said to be resonant and the formula for this is the same as for the series configuration, i.e. $f = \frac{1}{2\pi\sqrt{LC}}$. At resonance in a parallel circuit the currents in L and C will be equal and opposite, cancelling one another out, so the net flow will be zero. Therefore, with no current flowing between A and B the circuit has an infinitely high impedance since $Z = \frac{E}{I} = \frac{E}{O} =$ infinity.

In practice there is always resistance in a tuned circuit as well as pure inductance and capacitance. The coil itself will have the ohmic resistance of its windings. For this reason the circuit impedance will not be infinite, but will behave as if it has a resistance whose value can be found from the formula $\frac{L}{CR}$, where R is the resistance of the coil windings.

This dynamic resistance, as it is known, can be made large by increasing the value of inductance L compared with the capacitance C, or by reducing the resistance of the coil windings. Because of the high impedance at resonance, the parallel tuned circuit is often known as a rejector circuit. The only current flowing in this condition, between A and B, will

59

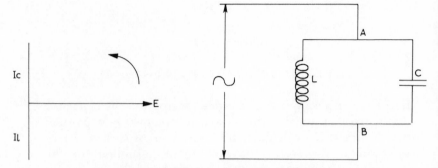

Fig. 25. Parallel circuit and vector diagram

be that determined by the dynamic resistance. The remaining current constitutes what may be termed a circulating current within the two branches of the circuit. This important fact will be discussed later in dealing with oscillators.

The ratio of the voltage across the resistance of a coil in a circuit to the voltage across the induction of the coil is also very important. It determines the 'goodness' or magnification factor called 'Q' of the tuned circuit, and a theoretical coil with zero resistance would have an infinitely high impedance.

In practice, the ratio of reactance to resistance is given by the formula $Q = \frac{2\pi f L}{R}$ which is sometimes written $\omega \frac{L}{R}$, where ω is $2\pi f$. This combination $2\pi f$ occurs frequently in formulae, and the form ω is often used to shorten the equation or formulae. Depending on conduction and materials, the Q of a coil will be about 50 to 400.

A graph may be drawn of the response either of a parallel-tuned or series-tuned circuit, shown in Fig. 26. As with the sine wave, the graph gives us a picture with information that can be taken in at a glance. The shape of the curve is important here since it gives information about the range of frequencies to which the circuit is responding. As mentioned in an earlier chapter, in radio terms we refer to a range of frequencies as a 'band' of frequencies and so the curves in Fig. 26 are showing the bandwidth of the circuits to which they refer. Tuned circuits can be used to identify specific frequencies, and then to reject or pass them. This is known as

60

SERIES CIRCUIT ACCEPTOR

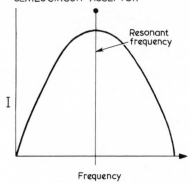

Resonant frequency

I

Frequency

PARALLEL CIRCUIT REJECTOR

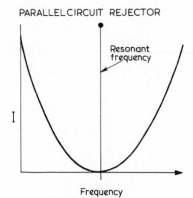

Resonant frequency

I

Frequency

Fig. 26. Graphs of series and parallel resonant circuits

filtering. Filters are used a great deal in radio and transmitter design, both in the rejection of unwanted frequencies and the enhancement of wanted ones.

In Fig. 27 a simplified arrangement of filtering is shown. A radio receiver RX is suffering from interference at a particular frequency. Once the interfering frequency has been established tuned circuits can be made up with the appropriate values of C and L, one series and one parallel. The parallel circuit A, being a rejector at this frequency, will impede the flow of signal current towards the receiver. At other wanted frequencies the circuit will offer either lower capacitive reactance at higher frequencies, or lower inductive reactance at lower frequencies. In the series circuit B, an acceptor circuit, these same interfering signals are offered a low impedance path to earth where they become neutralised. Again frequencies different from the interfering one are opposed by the high reactance of either inductance or capacitance.

Three important filter configurations are the low pass, high pass, and band pass filters (Fig. 28). The low pass filter has a series of inductances with capacitors forming a T-junction to earth, between 3 and 5 of these sections would be formed in a typical filter. The graph shows that up to the point marked cut off, the amplitude of the signal passed is constant, and at the cut-off point the signal is rapidly attenuated.

61

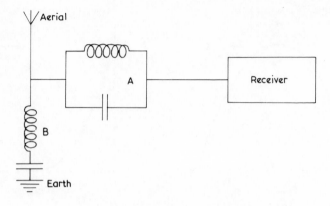

Fig. 27. Simplified interference filter

The frequency at which the roll off occurs, and the steepness of it will depend on the value of the components and their quantity. Up to a point the more stages a filter has, the more effective it is. As we have seen earlier, the reactance of an inductor increases with frequency, and that of the capacitor decreases, so there is a natural tendency for the high frequencies to be directed to earth in the low pass filter.

The main benefit of a filter such as this is in transmitters, where it is important to prevent the radiation of harmonics, i.e. frequencies which are multiples of the radiated fundamental frequency. An LPF filter can be chosen so that the wanted radiation is within the filter pass range and the unwanted ones lie beyond it.

The high pass filter is seen in Fig. 28b and here the position of the capacitors and inductors have been changed. The capacitive reactance decreases with an increase in frequency, and the reactance of the inductors increases, so frequencies lower than the threshold point in the graph are attenuated, whilst frequencies higher than this are passed through the filter. Again the component values will determine the point at which this occurs.

The high pass filter would benefit a receiver which was suffering from interfering signals lower than the frequency to which it was tuned, such as a TV receiver picking up signals of a lower frequency from a nearby amateur transmission.

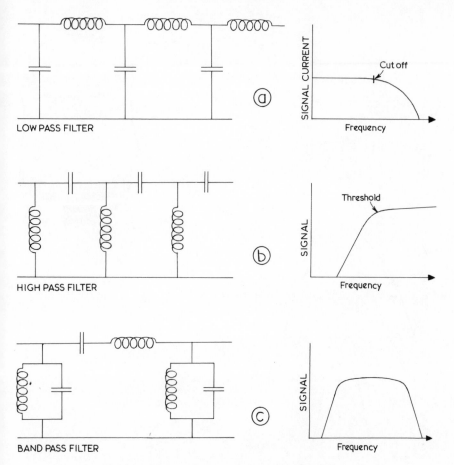

LOW PASS FILTER

HIGH PASS FILTER

BAND PASS FILTER

Fig. 28. Three important filter configurations

The HPF would attenuate the lower frequencies, but allow the wanted signals to pass through unimpeded.

The band pass filter, Fig. 28c, is really a combination of these two. Frequencies both above and below that for which the filter is designed are attenuated and only the band of frequencies between are passed through.

Coupled circuits are two tuned circuits so positioned that a magnetic coupling exists between the two inductors (Fig. 29). These circuits are used where the need is to select or amplify a given band of frequencies between the two stages. The

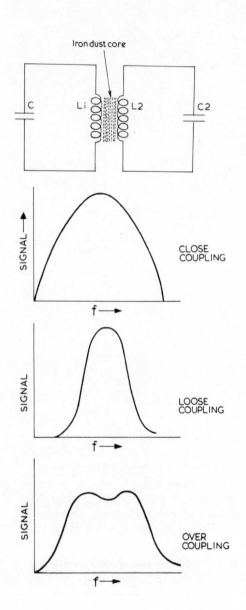

Iron dust core

C   Li   L2   C2

CLOSE COUPLING

LOOSE COUPLING

OVER COUPLING

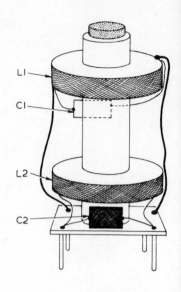

L1

C1

L2

C2

Fig. 29. Mutually coupled tuned circuit — IF transformer with three typical coupling characteristics

amount of coupling determines to a large extent the band-width, or selectivity of the coupled circuit and three examples are given in the diagram.

The most common form of mutually coupled tuned circuit is the intermediate frequency transformer of which two or more are found in many radio receivers, and invariably called the IF transformer for short.

The mutual inductance of two coils is also widely used in the form of a transformer. Basically this is induction in a secondary circuit by the alternating current flowing in the first or primary circuit. There is no tuned circuit in this case, and the purpose is usually to change, or transform, the voltage or current between one stage and the next, or to change the impedances between the two stages.

Transformers can deal with RF, AF, or LF (power) frequencies and the main difference will be in the number of turns and thickness of the conductors used in their construction.

Power transformers, as used in the supply to an amateur transmitter, may have to pass a current of 10 amps or more. In order to minimise any loss due to the resistance of the windings, they will be of large diameter and consequently bulky and heavy. On the other hand, a transformer coupling the RF signal from an aerial circuit to the following stage in a receiver, will be wound with very fine wire since the current it is carrying is only a few micro-amps.

Transformers are often on an iron core to enhance the magnetic coupling and to improve the efficiency of power transfer from primary to secondary. In power transformers the iron core is usually in the form of laminations to avoid the losses incurred when eddy currents are set up in the core material. The core acts like a one-turn secondary, and has circulating current induced in it which would result in a loss of power to the secondary. Laminating the core reduces these eddy currents to small proportions.

All transformers have losses and this must be allowed for in their design and use. For the purpose of explaining the transformer action these losses are ignored.

In Fig. 30, if the secondary has the same number of turns as the primary, the transformer has a turns ratio of 1:1 and

c

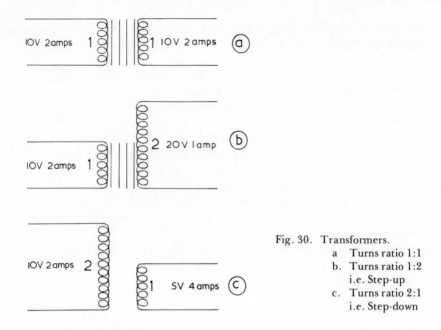

Fig. 30. Transformers.
a   Turns ratio 1:1
b   Turns ratio 1:2
    i.e. Step-up
c   Turns ratio 2:1
    i.e. Step-down

the same AC voltage current will appear at the output as at the input (ignoring losses). The second transformer has twice as many turns in its secondary as in its primary; it is a step-up transformer with a turns ratio of 1:2. The voltage of the output will be twice that of the input, but since power cannot be increased by a transformer the current will be halved. This is an important point to remember that the product of voltage and current at the secondary cannot exceed the product of these two in the primary.

In the example of the step-up transformer, the primary voltage is ten volts and the current 2 amps. Thus $10 \times 2 = 20$ Watts. In the secondary, the voltage is 20V and the current 1 amp: $20 \times 1 = 20$ Watts. The answer is therefore the same in both instances, so the condition has been satisfied. Commercial transformers are often rated in this way.

Many practical transformers have more than one secondary circuit. There may, for example, be a requirement in a transmitter for three different ranges of voltage and current. These can be obtained by having three secondary windings

66

on a transformer whose primary is connected to the current supply. The power rule, however, still applies: the total power dissipated in the secondary cannot exceed that delivered into the primary circuit.

The application of transformers in power supply circuits will be dealt with later.

Chapter Five

SEMICONDUCTORS

The day of the valve has passed and many will mourn its passing. For forty years it reigned supreme, so that many amateurs were brought up on valve operated equipment, often making their first amateur contacts on simple '3 + 1' sets. (Three valves and a rectifier).

The valve is basically a device for amplifying a signal, and has been almost completely replaced by the transistor. This device is hundreds of times smaller, more efficient in the consumption of current and cheaper to produce in large quantities. Compared with the valve, it needs only very small voltages — as low as 1˙5 volts — to make it conduct, and under normal operating conditions one can expect a long life without the gradual deterioration associated with vacuum valves.

The 'Cat's Whisker' crystal set was a very primitive type of radio receiver which depended for its function on the operator locating — by trial and error — the sensitive part of the crystal, so causing the receiver to operate. In this condition the 'Cat's Whisker' (the wire which was used to probe the sensitive area), together with the crystal, constituted an early form of solid state device.

Solid state, silicon chip, microprocessor, diode transistor, and integrated circuit (IC), are all names given to devices which depend for their operation on a group of materials called SEMICONDUCTORS.

In the periodic table, elements are grouped according to their characteristics, and group 4 is the semiconductor group whose outer shell is only half full of electrons. The maximum number of electrons which can orbit in the outer shell of an atom is 8 whereas the group 4 elements have only four. Two such elements are silicon and germanium. Both of these elements form themselves into a crystal lattice structure naturally, this occurs because each atom forms a co-valent bond with its neighbour. *See* Fig. 31a. In this simplified

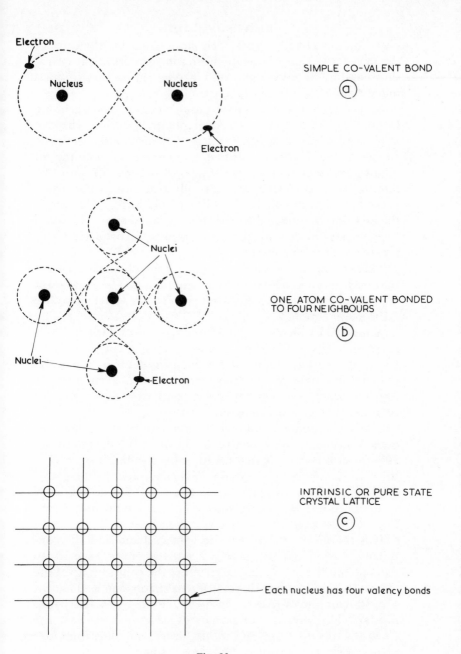

SIMPLE CO-VALENT BOND

(a)

ONE ATOM CO-VALENT BONDED
TO FOUR NEIGHBOURS

(b)

INTRINSIC OR PURE STATE
CRYSTAL LATTICE

(c)

Electron

Nucleus   Nucleus

Electron

Nuclei

Nuclei   Electron

Each nucleus has four valency bonds

Fig. 31.

diagram, one electron from each atom is sharing the orbit of the other in a figure of eight formation. In chemical terms, this bonding is very tight and causes the atoms to hold together in a rigid formation.

In fact each of the atoms has four electrons and each one forms a co-valent bond with a neighbour. *See* Fig. 31b. The outer shell of an atom is termed the valency band, hence the co-valent bond. Because of this symmetrical bonding of atoms, the structure of the material builds up into the familiar crystal formation. To illustrate this, the usual procedure is to draw a lattice like the one in Fig. 31c, where the bonding is indicated by the links between atoms. It would be confusing to develop the previous diagram and try to indicate each orbiting electron.

This crystal lattice diagram indicates the important feature that each atom is linked to four neighbours. The diagram is of necessity two-dimensional, whereas the crystal in practice is three-dimensional.

A material which is a good conductor is one which has a large quantity of drifting electrons temporarily separated from their parent atoms, and a good insulator is a material that has very few. Silicon and germanium, in their pure, or intrinsic state, are neither good conductors nor good insulators. Hence the name semiconductor.

The approximate resistance of germanium is 45 ohms per cubic centimetre, whereas that of silicon is 300,000 ohms per cubic centimetre. It is helpful in understanding this to consider an Energy Band Diagram. It is convenient for diagrammatic purposes to consider the electron that has drifted away from its atom, as having moved into a conduction band. Fig. 32 shows a segment of an electron in orbit.

It is the distance between the valency band and the conduction band, and the energy needed to move an electron from one to the other, that determines the conductivity of the material. Fig. 32 shows three different energy bands. That for a conductor shows that at ambient temperatures the bands overlap. This means that no energy is needed to cause the electrons to pass from one to the other and thus there are plenty of electrons available for conducting.

The next diagram shows the semiconductor situation, and

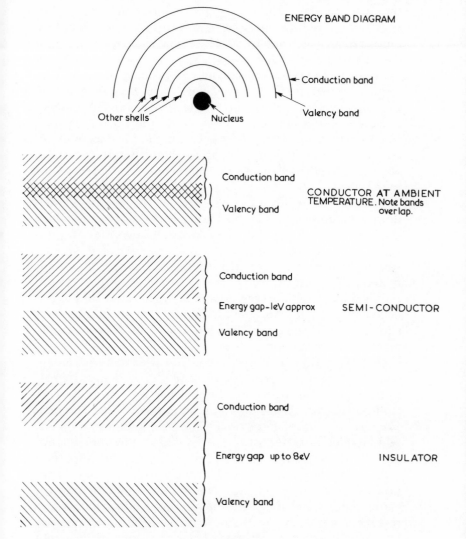

ENERGY BAND DIAGRAM

Conduction band

Valency band

Other shells   Nucleus

Conduction band

Valency band

CONDUCTOR **AT AMBIENT** TEMPERATURE. Note bands overlap.

Conduction band

Energy gap-1eV approx

Valency band

SEMI-CONDUCTOR

Conduction band

Energy gap up to 8eV

Valency band

INSULATOR

Fig. 32.

here there is a small energy gap to be overcome before electrons pass over into the conduction band. This means that under normal circumstances the conduction is very small compared with materials in the previous group. Finally, in

71

the third diagram, the energy gap is very much wider. The energy required to move electrons across these bands is measured in electrons volts. These materials then are poor conductors or good insulators, since there will be very few electrons in the conduction band except under great voltage stress. Pure germanium or silicon has a resistance of approximately 45 Ω per cc and 300 kΩ per cc, and if a potential difference is applied across a portion of such a crystal a small current will flow. This is called the pure state or intrinsic current. It is important to remember this current, which is the current due to the characteristic property of the pure crystal.

The key to semiconductors that we use in the various devices mentioned at the beginning is the manner in which the basic crystals are altered. Very small quantities of a different element are introduced in a process called doping, which significantly alters the conduction properties of the material.

When a crystal has been doped, the material introduced is called an impurity, and the crystal has become an impure or extrinsic semiconductor. There are two types, N-type, and P-type. Consider the N-type first.

In doping, tiny quantities of a pentavalent impurity are introduced into a very pure crystal. The process is an extremely exacting one, with the quantities being very carefully controlled. A pentavalent impurity is an element whose atom has five electrons in its outer shell. Two examples are antimony and arsenic. They are, of course, of different atomic weights; one has 51 electrons in total and the other 33. The important fact is that they finish up with five electrons orbiting in the outer shell or valency band. The impure atoms fit into the crystal lattice and link up with the surrounding atoms in a regular manner, except that one of the orbiting electrons is left spare without a co-valent bond thus creating a donor atom. This is indicated in Fig. 33a. The spare electrons thus introduced become available for conduction in the presence of an external PD (Potential Difference), and are called current carriers. Doping is at least ten times the intrinsic (pure) current level, which makes doped germanium approximately 4·5Ω/cc. N-type comes from the polarity of the introduced current carrier, which being an electron is negative.

P-type semiconductors are made by introducing a trivalent impurity, in which atoms have three orbiting electrons in their outer shell. Examples are indium or gallium. Again, introduced in very small quantities they link up in the lattice network, but this time one of the links is missing, creating a gap in the lattice at that point. This gap is known as a 'hole', *see* Fig. 33b. It is called P-type material because the hole can accept one electron and by so doing is opposite in polarity to the N-type carriers, i.e. positive.

Conduction in N-type materials is by electrons moving towards positive, and in P-types by holes moving towards negative. The concept of a hole behaving as a current carrier is difficult to grasp, but is very important to the understanding of the function of semiconductors. In P-type materials there is a predominance of positive charges and in N-type materials of negative charges.

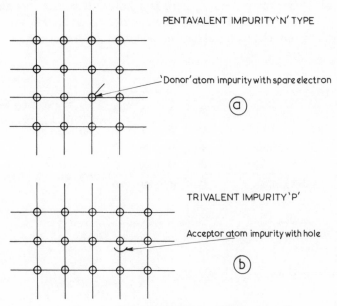

PENTAVALENT IMPURITY 'N' TYPE

'Donor' atom impurity with spare electron

ⓐ

TRIVALENT IMPURITY 'P'

Acceptor atom impurity with hole

ⓑ

Fig. 33. Doped semiconductor material

C*

An old explanation of a hole movement, but an effective one, is the doctor's waiting room, ten patients are sitting in ten chairs along the waiting room wall. The doctor calls the first one in, thus leaving an empty chair (hole) nearest to the doctor's door. The next patient moves up into the chair just vacated and the patient next in line moves up one too. The empty chair moves down the line until it is furthermost from the doctor's door. In fact each patient has moved one position while the hole has moved ten; so it is with semiconductors.

The PN Junction. The simplest device is the PN junction, creating a single lattice structure with P-type in one half and N-type in the other. It is important that the junction is part of one continuous structure, and not two separate lumps stuck together. There are many different processes for achieving this, such as alloying and diffusing, which result in the junction occurring within the crystal structure. In Fig. 34. a PN junction is shown with symbols indicating the extrinsic (doped) current carriers only. An ionised atom is one which has a charge either positive or negative, due to the imbalance of its electrons. In the crystal lattice structure where a trivalent impurity occurs, there is a hole due to the fact that the impurity has only three valency electrons. If this hole is filled by an electron the atom becomes ionised, it has one

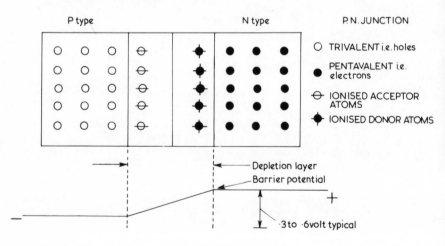

Fig. 34. PN junction

electron more than it should and therefore assumes a negative charge. When a pentavalent impurity with a surplus electron loses that electron it assumes a positive charge.

When the PN junction is created there will be some recombinations in the region of the junction. Electrons in the N region will drift across the junction, attracted by the holes in the P region. This leaves donor ions in the N-type material and acceptor ions in the P-type material both uncompensated; that is they are positive and negatively charged respectively. This apparent potential difference prevents any further diffusion from taking place, since the remaining electrons in the N region are opposed by a barrier of negatively charged ions in the P region. This is known as the *barrier potential*. The area where these reactions have taken place is however known as the depletion layer, since it has been depleted of current carriers. The thickness of the depletion layer is dependent upon the amount of doping, but it is extremely small, a few millionths of an inch, which helps to explain why many semiconductors are so sensitive to misuse or overloading.

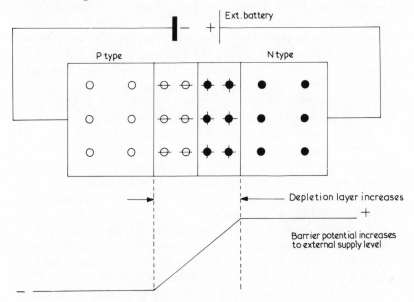

Fig. 35. PN junction. Reverse bias

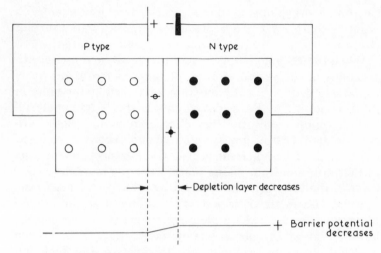

Fig. 36. PN junction. Forward bias

PN Junction Biasing. So far the PN junction has been considered in isolation, the PD across its junction occurs
naturally when the device is formed. Biasing is the term given
to the application of an external PD from say a battery.
Reverse bias is that which is applied to the PN junction so
that the applied PD is aiding the barrier potential. The effect
will be to increase the depletion layer (Fig. 35). The only
current to flow will be the very small leakage current (the intrinsic current), the current which would have flowed in a
piece of pure crystal without the impurities added.

Silicon and germanium both possess the property of
decreasing resistance with increasing temperature. This is
because external heat causes a greater number of electrons to
overcome the energy gap between valency band and conductor band (Fig 32). Leakage current will therefore be
determined largely by temperature, and the area of the junction. The reverse or leakage current is usually constant at a
given temperature regardless of the change in reverse bias.
This is because the current is saturated, that is to say all the
intrinsic current carriers available are contributing to the
reverse current. Temperature will however increase leakage
current, an important point with all semiconductors. There is

76

a point with all PN junctions, called the breakdown voltage where reverse current will increase rapidly.

Forward bias is where the external PD is connected in op-position to the barrier potential (Fig. 36). The depletion layer is reduced and this allows a current to flow which will vary in proportion to the applied PD.

The PN junction forms the device we know as the DIODE. There are a great variety of diode types, with many different characteristics, ranging from ones which handle tiny currents of a few micro-amps, to power devices which are capable of passing several amps. The diode is undirectional, that is to

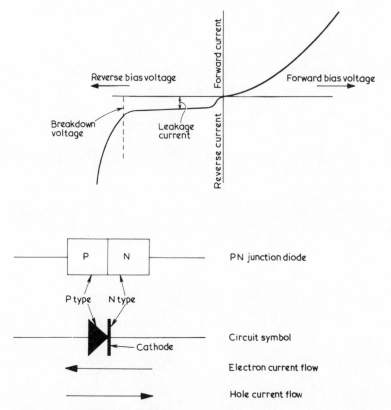

Fig. 37. The PN junction diode with graph showing forward current flow and reverse leakage current flow

say it has a high resistance to current flow in one direction. The characteristics are shown in graph form in Fig. 37. One of the functions of a diode in an electronic circuit is to convert an alternating current into a uni-directional current. The process is called rectification, and we shall see later how it is applied to power supplies and detector circuits. Fig. 38 shows the basic process together with the symbol used for diodes. It can be seen that the diode conducts when it is forward biased on the positive half cycle, but does not conduct on the negative half cycles, where it becomes reverse biased. The circuit would work equally well if the diode were reversed, except now the negative half cycles would appear as the output.

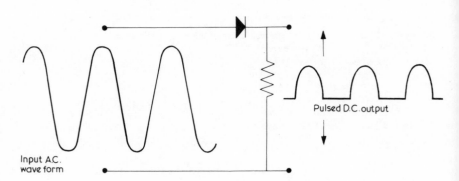

Input A.C.
wave form

Pulsed D.C. output

Fig. 38.   The diode used as a rectifier

Chapter Six

TRANSISTORS—Part I

The bipolar transistor consists of a sandwich of·PN materials, or two PN junctions back to back. There are obviously two ways in which this can be done, either p-n-p or n-p-n, *see* Fig. 39. Both are utilised and in fact are distinguished by reference to their type as *p-n-p* or *n-p-n*. At each junction of the n-p-n transistor there will be a depletion layer and a barrier potential as before. In constructing a transistor, however, the base region, as the centre portion of the sandwich is called, is only lightly doped.

If an external PD is applied so that the emitter to base junction is forward biased, this part of the device will be low resistance. If at the same time the base to collector junction is reversed biased then this section will be high resistance. It is usual to refer to the emitter base junction as the emitter junction, and to the base to collector junction as the collector junction.

In order to achieve this the left hand 'diode' (*see* diode equivalent circuit, Fig. 39) would need to have a positive voltage on its base connection compared to the emitter connection. The collector voltage would need to have a positive voltage on its collector compared with the base in order to achieve reverse bias.

Simply put, the transistor produces a gain because it is able to cause a similar current to pass through a low resistance and a high resistance. The reduction of the barrier potential at the emitter junction caused by the forward bias will allow many electrons to be emitted into the base region. The base is lightly doped, thus only a few holes exist for recombinations to take place (less than 2 per cent). The base region is also very thin and the surplus electrons drift into contact with the collector junction and are taken into the collector region attracted by the positive potential to form the collector current.

The input resistance is low, but the output resistance is

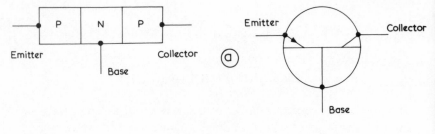

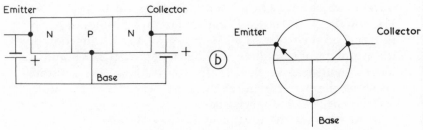

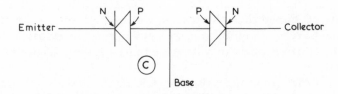

Fig. 39. a. The PNP transistor
          b. The NPN transistor
          c. The diode equivalent of b

high, with almost the same current flowing. Thus a gain in power exists, since the formula for power is $I^2 \times R$. $I^2 \times R$ (output) will be greater than $I^2 \times$ little R (input).

The name transistor comes from the description *transformed resistance*.

The action of p-n-p transistor is broadly the same. The main difference is obviously the reversal of the voltage polarities required to bias the two junctions, since these are now the opposite way round.

Less obvious is the fact that in n-p-n types the conduction is mainly by electrons—the majority carriers. Whereas in the p-n-p type, conduction is mainly by holes—the majority carriers.

80

The amount of current which flows in the collector region of either type, is dependent on the number of current carriers attracted into the base region in the first place. Since the emitter junction is low resistance, a very small change in the forward bias voltage will cause a comparatively large change in the emitter current and therefore the collector current too. This use of a small change at the base, to control a large change at the collector is the basis of the transistor as an amplifying device.

In order to control the output of a transistor, a load is required in the collector circuit, this may take the form of a resistor, or it could be the impedance of an inductor or tuned circuit.

Before we look at the way this works, let us examine a simple way of providing bias from a single source, and at the same time look at the transistor in its most usual configuration, 'common emitter', Fig. 40a. So called because the emitter potential is constant and common to both collector and base, whereas in the previous example, 'common base', the base potential is constant and common to both emitter and collector. As can be seen, the base current that flows through R1 will cause a voltage drop across it. According to Ohm's Law, this leaves a small base voltage positive with respect to the emitter and therefore forward biased, while the remaining voltage appears as reverse bias across the collector junction.

The arrangement is simple and will work, but a better form of biasing is shown in Fig. 40b. Here an extra resistor R2 has been introduced which forms a potential divider, this arrangement is more stable, because the bias voltage is now much less dependent on the base current, which, if it fluctuates, causes the bias voltage to fluctuate. This is largely overcome by making the bias voltage independent of the base current in the divider network shown.

The current flowing through R1 and R2 will depend on their total resistance and the voltage at b will depend on their ratio to each other. Actual calculations will in practice be more complex than this because the resistors are in parallel with the resistance of the transistor itself.

Let us now look at a common emitter configuration, and

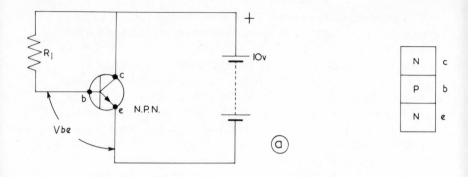

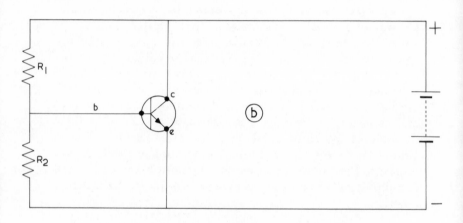

Fig. 40. a. Simple bias arrangement
b. A better arrangement — bias by potential divider

how its output can be controlled. Fig. 41 shows an n-p-n transistor forward biased with R1 and R2 and a load resistor RL. The purpose of the capacitors C1 and C2 is to isolate one stage from the next, and to prevent voltages reaching an adjacent stage. Often called DC blocking capacitors, they are chosen to have a low reactance at the frequencies being amplified and can be considered as short circuits, as far as the signal is concerned. The purpose of Re will be considered in a moment.

82

If the input signal at the left of Fig. 41 is applied to the base via C1, it will cause the bias first to increase and then decrease by the amount of its peak value. These very small changes in the forward bias at the emitter junction will cause much larger current changes to occur in the collector region because of the transistor action. These same current changes will occur in the load resistor RL and also produce voltage changes, since Ohm's Law states that current is proportional to voltage when the resistance is constant.

When the input signal is positive-going, notice that the amplified output signal is negative-going. A phase inversion always takes place in a common emitter amplifier. This is because the top end of RL is at a fixed supply potential and therefore an increased voltage drop across it must result in the collector voltage decreasing.

The ratio of this small current change in the base to the current change in the collector is called the current gain of the device, β or h$_{FE}$.

The β of modern transistors varies enormously, typically from 20 to over 600. The ratio of voltage input to voltage output is termed h$_{RE}$. There are many parameters, or charac-

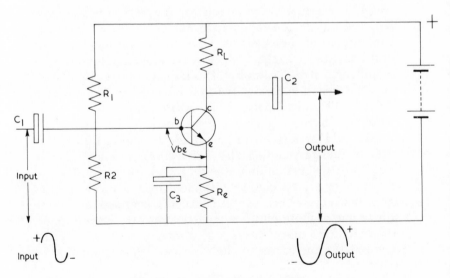

Fig. 41. Common emitter amplifier

teristics, of a transistor. These two and the following one are the most important. α is the ratio of how much emitter current gets into the collector region. α is always very close to unity and in the description of power gain made earlier the expression 'almost the same current in the emitter as in the collector' was used. In practice of course, some current would flow out at the base region, say 2 per cent, leaving 98 per cent to flow in the emitter; this would give a ratio or α of $\cdot98$.

The purpose of Re in Fig. 41 is to act as a stabilising component. As mentioned earlier, the semi-conductor resistance will fall as temperature increases. This will allow more current to flow which in turn creates more heat and less resistance. This effect can accelerate (a condition known as thermal runaway) until so much current flows that the device is destroyed. Re in Fig. 41 will have a voltage drop across it, proportional to the current flowing through it. This voltage drop will determine the potential at e, the emitter. The bias at the base emitter junction will be the difference in potential between the junction of R1 and R2 and the top of Re. If, due to thermal activity, current flow increases through the transistor, and therefore Re, then the voltage across it will increase. Therefore, as the voltage at C increases the difference Vbe will decrease, thus reducing the forward bias and therefore current, compensating for the thermal increase. The purpose of the capacitor C3 is to smooth out the short term changes in current which arise when the transistor is amplifying a signal, allowing it to respond only to the longer term changes due to the thermal effect. The capacitor in effect short circuits the signal voltages that appear at e.

Power transistors generate heat which limits their effectiveness and arrangements are made to conduct this heat away to the surrounding air. This consists of clamping the transistor to a surrounding area of metal, often with fins, so the heat generated can be dissipated over a much greater area. A structure such as this is called a heat sink and there will be one or more found in virtually every transistor circuit. Manufacturer's data sheets will always give details of the operating temperatures of their devices, which must not be exceeded.

Fig. 42 shows the third configuration of a transistor am-

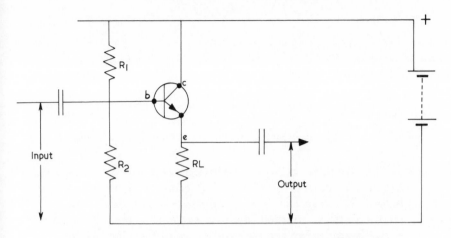

Fig. 42. Common collector configuration or emitter follower

plifier, namely the common collector, or emitter follower. In this arrangement the collector voltage remains constant while the input base voltage and output emitter voltage vary. There is no phase inversion with this circuit, the output follows the input in phase, hence emitter follower. The input impedance is high, whilst the output impedance is low and this is useful for matching different impedances in an amplifier, typically a high impedance microphone into a low impedance amplifier stage input.

It is also useful as a 'buffer' or isolating stage between one circuit and another where interaction between the two is undesirable. There is no voltage gain from the common collector, since the voltage at e must be less than that at b. There is however current and power gain.

The circuit in Fig. 41 can be considered as a basic AF small signal amplifier. For an amplifier at RF frequencies the load is usually in the form of a tuned circuit resonant at the frequency being amplified. The inductance being wound or tapped at a point which gives the impedance which matches the transistor. Because the circuit is resonant and therefore of high impedance at this frequency, it will produce an output across the load. At other frequencies, the circuit is low impedance and will produce little output.

85

Fig. 43. Distortion produced from over-driving amplifier in Fig. 41

The circuit is therefore selective as well and is the basis of the IF transformer used in radio receivers as we mentioned before. The transistor can only operate within the limits of its characteristics, marked by two extremes. If a very large signal is applied to the base of our amplifier, Fig. 41, on the positive going portion, with heavy forward bias, the junction current will increase causing a large voltage drop. The voltage drop cannot exceed the supply voltage so when the collector voltage has come down to zero, or very near it, the transistor has 'bottomed'.

Transistors are often used in this way as solid state switches. A sufficiently large voltage being used to bias the device 'on', i.e. so the collector bottoms.

When the negative portion of the large input signal occurs the base becomes negative with respect to the emitter, the forward bias is lost and the device is 'cut off'. Current will cease to flow. At this moment the collector voltage will be at, or near, the supply voltage. The resulting output will be something approaching a square wave, Fig. 43, which is a very distorted version of the input signal. Really the only factor which has been preserved is the frequency. This type of wave form (the square wave) can be shown by complex analysis to embrace a large range of harmonics. Again, this feature of the amplifier may be used for various purposes in electronics. However, if we are looking for linear amplification, that is an output which is similar in shape to the input, only magnified, these distortions have to be avoided.

Class A is the name given to the condition of operation that enables near linear amplification to take place. The amplifier circuit is redrawn as Fig. 44a, showing just the bias and the load components. R1 and R2 are chosen so that the

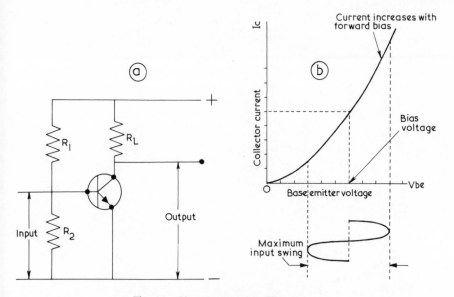

Fig. 44. Class A operation (linear)

amount of forward bias causes the transistor to conduct a current about midway between cut off and bottomed. Provided the input signal is not too large, and stays within the straight part of the curve in Fig. 44b the output will be reasonably linear. In class A collector current flows all the time, whether or not there is any signal input; also, we have seen that the input signal needs to be small to avoid distortion. Class A therefore imposes a constant current drain on the source of supply, which means it is inefficient where small portable batteries have to be used.

Class B is the name given to the operation in which the bias R1 and R2 is chosen so that it occurs at the point of cutoff, so in the absence of an input signal no current flows (Fig. 45a). It is obvious that this form of amplification will result in severe distortion, since the device will only conduct on the positive half cycle of the input. This type of amplifier will contain harmonics in its output. Two Class B amplifiers can be used in a configuration known as push-pull, so arranged that each transistor amplifies one half of the input signal, which then restores the original shape. *See* Fig. 45b. In this

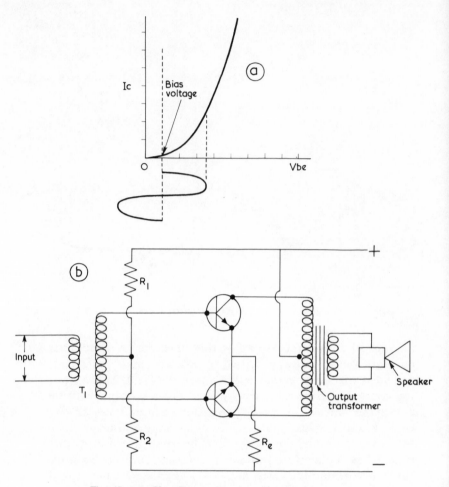

Fig. 45. a. Class B operation (non-linear)
          b. Push-pull arrangement for restoring linearity

basic circuit the input is via a centre tapped winding which provides opposite phases of the signal at each base. R1 and R2 are the base resistors, Re the stabilising resistor, and in this circuit the output load is the inductance of an audio transformer which is in turn coupled to a speaker. The input will induce in the secondary of T1 voltages which are alternately positive and negative at opposite ends of the winding. Thus one transistor will conduct while the other is cut off, and then

88

the opposite will take place. The two resulting half wave outputs will, in a properly biased circuit combine to give a complete waveform with no distortion.

In the type of operation, known as Class C, there is no bias applied to the base of the transistor and therefore the device will conduct only in the presence of a signal large enough to cause the emitter junction to become forward biased during the peak of its input (Fig. 46). As the transistor is only conducting during part of the signal cycle, the output is extremely distorted. It is in fact a series of pulses and rich in harmonics. The frequency of the signal is preserved and the pulsed output makes use of the greater part of the transistor voltage swing, and again draws no current in the absence of a signal input. The input signal for Class C has to be large enough to drive the device into forward bias and is usually preceded by a driver voltage amplifier. Used in RF amplifiers, the sinusoidal nature of the input signal is normally restored in a tuned circuit forming the collector load.

In the circuits discussed so far, the transistors have been of the n-p-n type. In general, all these circuits would behave in exactly the same way with p-n-p types, remembering that all the polarities of supply and bias would be reversed.

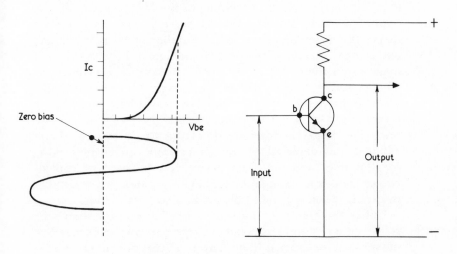

Fig. 46. Class C operation (non-linear)

Chapter Seven

TRANSISTORS — Part II

The transistors described so far are known as bipolar transistors, so called because conduction is by means of positive hole and negative electron current carriers. They also have a relatively low input impedance, due to the fact that current flows in the base circuit.

A semiconductor device developed later than the bipolar is the unipolar or Field Effect Transistor (FET), in which conduction depends on majority current carriers only, i.e. holes in P-type material and electrons in N-type material.

To understand how current flow and output are controlled refer to Fig. 47. Note that there are different names for the connections but the source corresponds to the emitter, the gate to the base and the drain to the collector. The N-type material, the channel, has a ring of P-type material diffused co-axially around it. This is the gate region.

The amount of current that will flow from source to drain depends on the amount of doping of the N-type material, the applied voltage, and also the diameter of the channel. The larger the diameter, the greater number of current carriers and hence current. The effective diameter of the N channel depends on the width of the depletion layer at the junction. This in turn is determined by the reverse bias. As this is increased the depletion layer increases, until being annular it meets at the centre. This point is called the pinch-off voltage since the depletion layer occupies the whole section of the channel and depleted of current carriers, no current flows from source to drain. The device can be used to control current and voltage in an external load by choosing a suitable reverse bias voltage and applying an input signal to it.

Unlike the bipolar transistor the FET has no flow of current at the gate region (except for leakage current) and therefore the input impedance is high. This is a great advantage in certain applications in radio and gives similar characteristics to

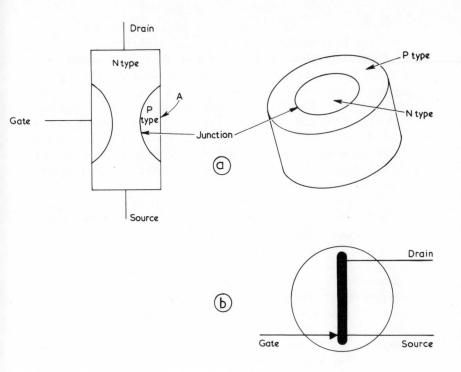

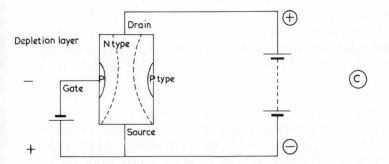

Fig. 47. Field effect transistor
    a. Showing sections revealing the P and N type materials
    b. FET circuit symbol for N channel (for P channel the arrow is reversed)
    c. Showing bias arrangement and depletion layer

the triode valve of old, but without the disadvantages of heat, size and high voltage.

Three other semiconductor devices likely to be encountered in radio circuits are the LED (Light Emitting Diode), the Zener diode and the Varactor. The LED is a junction which, when forward biased, gives a light output. Made from gallium phosphide it gives a useful source of light, red, green, or yellow, for indication purposes. The advantages of the LED over filament bulbs, for this purpose, are its low current drain, just a few milliamps, and low voltage, together with long life and durability.

The zener diode is used in voltage stabilisation circuits, and has an important place in amateur equipment, where stabilisation is an important factor.

As can be seen in Fig. 48 the zener is reverse-biased so that normally it would not be expected to conduct. In a diode very little current flows in the reverse bias mode; that which does is the leakage current. However, a condition is reached where a high reverse voltage will cause current to flow at the breakdown point. Current then increases very rapidly, this being known as the 'avalanche' effect. Specially constructed diodes (zeners) are made to operate in this region and without permanent damage. The reverse voltages at which this controlled breakdown occurs can vary from about $2 \cdot 5$ volts upwards. Fig. 48a shows a typical characteristic. Such diodes can be used to stabilise voltages in circuits where there is fairly small change in the load conditions, such as the oscillator, which will be discussed shortly. In the circuit shown in Fig. 48 a small variation of the input voltage will cause the breakdown current in the zener to vary, which will compensate for the changes in the current through R1 and the voltage at the output point will stabilise — in this example at about 7V. Zener diodes are available in fixed values ranging in useful steps to about 150 volts. Also available are programmable zeners where the breakdown voltage can be set to any desired value.

The third diode in Fig. 48d is the varactor. Any diode when reverse biased, consists of two conducting regions, the P- and N-types, separated by an area of insulation which is the depletion layer where little current flows. This of course is

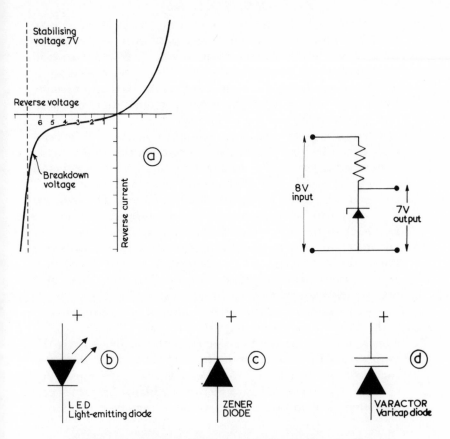

Fig. 48. a. Zener diode characteristics and stabilising circuit
b. Symbol for LED, light emitting diode
c. Symbol for zener diode
d. Symbol for varactor diode

the definition of a capacitor. Any diode will have capacitance of this nature, which may need to be allowed for in its application. In the varactor this effect is made use of and the device becomes a variable capacitor whose capacitance is inversely proportional to the applied reverse bias.

As the reverse bias voltage is increased, the depletion layer becomes wider and the capacitance less, within fairly narrow limits.

93

Oscillators. If an amplifier with a tuned circuit as its load receives a little of the output at its input in the correct phase, a sinusoidal signal will be maintained. The frequency will depend on the resonance of the tuned load. Such a circuit is said to be oscillating. The oscillator, both fixed and variable frequency, plays a very important part in radio, in transmitters, receivers, and test equipment.

There is a great variety of oscillator circuits, and in principle they all work by arranging for some of the output signal to have sufficient positive feedback to the input for oscillation to be maintained. Fig. 49 shows one form of basic oscillator circuit in which the basic signal path is shown. The resonant frequency will be determined by the values of L1 and C1. Feedback to the input is via L2, the correct phase being obtained by the polarity of L2 windings. Note that in order for the signal to be phase aiding, there must be 180^0 phase shift at the emitter to compensate for the inversion that takes place in the transistor.

Another circuit likely to be encountered in receiver circuits is the Colpitts oscillator, named after its originator. In Fig. 50a, again showing the basic signal path, feedback is via the tapped tuned circuit at the junction of C2 and C3, these two capacitors with L1 are the frequency determining components. By making C1 variable, the oscillator frequency can be varied. It is in fact a VFO, variable frequency oscillator;

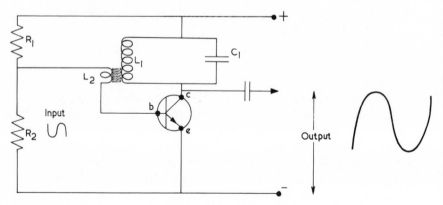

Fig. 49. Simple RF oscillator

note that this is a common collector circuit, and therefore the input signal is in phase with the output signal.

Quartz Crystal. Another type of oscillator is that which is controlled by a quartz crystal. To understand how this very stable form of oscillator works, a brief description of the piezo-electric effect is required. Quartz and certain other crystals exhibit the property of producing a charge across their surface when subjected to mechanical pressure. Conversely, when an electrical potential is applied across the same surfaces a stress is set up in the crystal. If the crystal is cut in a special manner, and formed into thin slices, the stress set up by an applied voltage can cause it to vibrate. The resonant frequency of any crystal will depend on the dimensions, mainly the thickness.

Crystals can be cut so that they vibrate mechanically at radio frequency up to 10 MHz and beyond. These vibrations are very stable and can be made very accurately. Activated by voltage, oscillation of the crystal can be maintained and used in the same way as a tuned circuit. Indeed the quartz crystal can be considered as equivalent to a very stable, highly accurate tuned circuit, and as such can be used in oscillator circuits. Such a circuit would have a very high Q. Quartz controlled oscillators are useful where simple circuits at fixed frequency and great accuracy are required. Many clocks and watches owe their accuracy to a quartz crystal oscillator. A simple circuit and the crystal symbol are shown in Fig. 50b.

As in most of the symbols, the one for a crystal reasonably depicts the actual construction. The thin wafer of crystal is held between two highly polished plates from which the external electrical connections are taken. The plates are spring loaded, with sufficient tension to hold the crystal, yet allow it to vibrate. Most modern crystals are sealed inside metal cases, with the connections extended to two pins protruding from the base. A matching base can be soldered permanently into the circuit making it simple to change frequency by plugging in the appropriate crystal. Small changes in the frequency of crystal oscillation can be effected by arranging for a few pico farads of variable capacitance to be in parallel with the crystal, often called a trimmer.

95

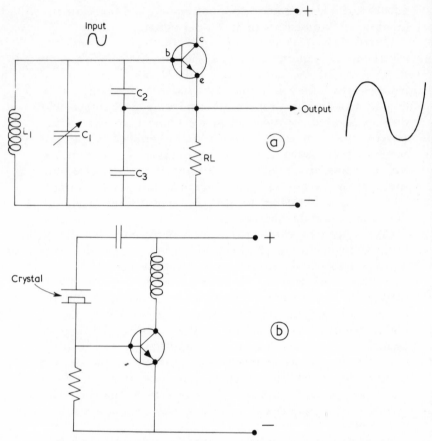

Fig. 50. a. Colpitt's oscillator
          b. Crystal oscillator — note the circuit symbol for quartz crystal

Frequency Changing. It is often desirable and sometimes unavoidable in transmitters and receivers to change the frequency in one stage of the circuit to some different frequency in the next. We shall see in later sections where this is necessary. The two basic ways of changing frequency are by the frequency multiplier, and by heterodyning. First, the frequency multiplier. This is simply an amplifier with a resonant collector load which is tuned to a multiple or harmonic of the fundamental frequency at its input. In practice, useful outputs are seldom achieved above the fourth harmonic. Therefore if a multiplication factor of say six were required, it would be better to achieve it in two different stages: a three times, followed by a two times. The frequency multiplier is a simple arrangement but the change in frequency that can be achieved is restricted to multiples of the fundamental.

The process which can produce a new frequency, not harmonically associated, is heterodyning. Here the signal to be changed (f1) is fed into a transistor which acts as a mixer, and also fed into the mixer is the output from another transistor (f2) acting as an oscillator. When this is done, the output from the mixer will consist of four components: f1, f2, and f1+f2, and f1−f2. This is illustrated diagrammatically in Fig. 51. For example, by mixing 1500kHz with the 1000 kHz frequency, the output in addition to these original frequencies, also contains their sum and difference. One of these two new frequencies can be selected by a resonant tuned circuit which will reject or attenuate the other three. Using this principle, it is possible to produce any frequency from the f1 input by choosing the appropriate f2 oscillator frequency to mix with it.

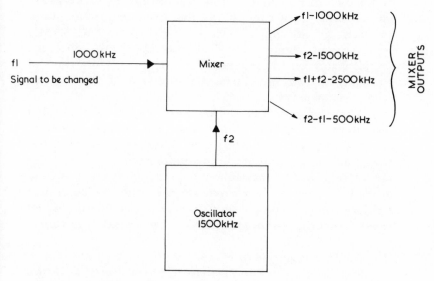

Fig. 51. Block diagram of frequency changer

Chapter Eight

MODULATION

Modulation, in radio terms, is the process of encoding audible frequency information on to the radio frequency, which is usually referred to as the carrier frequency, or simply the carrier. The earliest and simplest method of conveying intelligence from transmitter to receiver was by means of the Morse code. This form of communication was known as telegraphy and was already in use before radio itself was developed. Here the letters of the alphabet, the numerals, and certain other signs are represented by various combinations of dots and dashes. These characters would be sent out by a Morse key which would cause the make-and-break of an electric current to be conveyed along telegraph wires between the two points of communication. Developed in 1838 by an American, Samuel Morse, this code is still very much in use. It made the change from wire to 'wireless' quite easily and is used in its simplest form to interrupt the carrier wave into bursts of Morse code characters. In amateur radio, Morse transmissions are almost always referred to as CW. Strictly speaking, this should be called interrupted CW, since the initials stand for Continuous Wave. In the classification of emissions listed in the appendix to the amateur licence, this type of emission is listed as A1.

The transmission of the spoken word is a more complicated procedure, though far simpler from an operating point of view since no code has to be learnt. It can be achieved in two ways, either by amplitude modulation or by frequency modulation. In both cases the unmodulated signal consists of a steady frequency, the carrier, varying neither in amplitude nor frequency. In simple terms the audio modulating signal causes either the amplitude to vary or the frequency to vary.

Amplitude Modulation (AM)

The radio frequency or carrier can be considered as f1 and the AF modulating signal as a steady tone f2. These two signals are to be mixed together and the result of this mixing will be four frequencies f1, f2, f1+f2, and f1−f2. As a typical example assume the carrier (f1) to be 1·9 MHz, a frequency at the centre of the lowest of the amateur bands, and the AF tone to be 1·0 kHz (1,ooo Hz). This frequency is comfortably within the audio range. The resulting modulation will give: —

f1	=	1·9 MHz or 1,900 kHz
f2	=	1,000 Hz or 1 kHz
f1+f2	=	1901 kHz
f1−f2	=	1899 kHz

All but the second of these are now radio frequencies, and thus capable of being propagated. In an AM transmitter all three separate frequencies would be radiated. In practice f2 is likely to be a band of frequencies about 100 Hz to 4·0 kHz. This produces two bands of frequencies, one higher than the carrier and one lower, referred to as the upper and lower sidebands. This is illustrated graphically at Fig. 52a. Note that the total bandwidth occupied by the modulated signal is twice that of the audio band, i.e. 8 kHz. A receiver tuned to the carrier frequency will normally receive all three signals more or less equally. These three frequencies (considering for the moment the single tone modulating frequency) are combined to produce the wave form shown at Fig. 52b. This is produced in a rather complex fashion, but in fact the carrier f1 stays at a constant frequency and amplitude. F1−f2 is slightly lower in frequency and f1+f2 is slightly higher. These two sidebands are continuously changing their phase relationship with the carrier, sometimes in phase and adding to the amplitude, sometimes out of phase and diminishing it.

The result of this is the AM envelope wave form, whose period of rise and fall is that of the modulating frequency. When the peaks of modulation just meet, as in Fig. 52c the carrier is said to be 100 per cent modulated. Any increase of power in the audio signal at this condition results in over modulation.

This results in distortion at the receiving end and the break up of the carrier which produces spurious sidebands. These

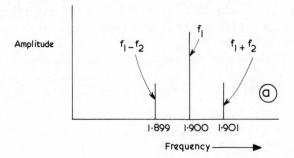

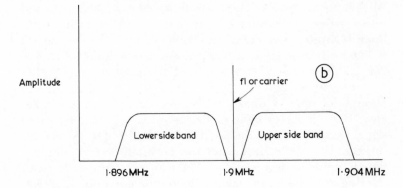

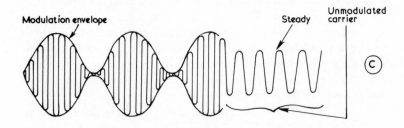

Fig. 52. Amplitude modulation
   a. Carrier modulated with single tone of 1 kHz
   b. Carrier modulated with band of frequencies 100 Hz –
      4 kHz
   c. Resulting 'envelope' waveform

spread further than the normal bandwidth of the signal and as a result may cause interference to other amateurs and other services. This type of emission is classed A3E.

For communication purposes AM has disadvantages. Firstly it is inefficient in power terms and secondly the bandwidth occupied is twice that of the audio frequency being modulated. Both of these factors are important to the amateur because he has a limited power with which he is allowed to operate, and therefore needs to make maximum use of it. Also the bands are now very crowded and a minimum amount of bandwidth needs to be used in order to accommodate as many channels as possible.

Single sideband emissions (SSB) achieve both of these objectives and this form of amplitude modulation is in widespread use throughout amateur radio.

The audio information required to be sent occurs in both upper and lower sidebands both are the difference between carrier and modulating frequency. The carrier itself contains no audio information, and therefore the receiver is extracting information which is carried in one sideband. The AF power required to modulate fully an AM transmission is about 50 per cent of the carrier power. This means that for a carrier of 100 watts power some 50 watts of audio power is needed for 100 per cent modulation. Of this 50 watts, 25 watts would be in each sideband so at the receiver the information extracted is that contained in one sideband, and produced with a power of 25 watts. The remaining sideband and carrier, produced by a power of 125 watts, is wasted. If only one sideband were to be radiated, its strength could be increased by five times using no more power than before. One sideband would of course also halve the bandwidth of the AM emission.

In essence, this is SSB. Within the transmitter AF and RF frequencies are mixed together at low levels of power. One of the sidebands is eliminated together with the carrier, and the remaining sideband is amplified in a linear manner to the level of power required for radiation.

Although simple in principle, the generation of SSB signals and their reception requires more complexity in the transmitter and the receiver than is the case with AM. Nevertheless, this form of emission was developed and pioneered by amateurs, as were many other aspects of radio. This class of emission is known as J3E.

Frequency Modulation

This form of modulation causes the frequency of the carrier to vary at the rate of the modulation of the AF signal, the amplitude of the carrier remaining fixed. The amount by which the frequency changes, known as the deviation, is proportional to the amplitude of the modulating signal. Compared with the broadcast FM signals, amateur FM deviation is very small. NBFM (Narrow Band Frequency Modulation) deviation is of the order of $2 \cdot 5$ kHz, that is to say the unmodulated FM carrier will increase by $2 \cdot 5$ kHz for full modulation.

The modulation process is very simple and easily achieved at low power levels, and lends itself readily to use in mobile and portable transmitters for these reasons. No particular difficulties are introduced at the receiving end and small FM transceivers have become very popular equipment on the amateur scene in recent years. Due to the frequency-varying nature of the modulation rather than the varying amplitude, over modulation and the creation of interference is greatly reduced in this mode.

Chapter Nine

RECEIVERS — Part I

The receiver could well be described as the heart of the amateur radio station, and it must be capable of performing three basic functions. One, it must be able to select one desired signal from among the thousands which surround it at any given moment. Two, it must amplify the tiny voltage generated by the signal, which is perhaps only a few millionths of a volt, to a level which will produce sufficient power to drive headphones or a loudspeaker. Three, it has to extract the modulated AF information. This last process is known as demodulation, or detection.

In the early days of radio, a very simple form of receiver was employed. This was known as a crystal set or 'Cat's Whisker'. The device was basic in the extreme and often constructed by the listener himself. The names were derived from the fact that the essential part of the receiver — the detector — involved finding the sensitive part of a piece of quartz crystal by means of a probe made of fine wire — the 'whisker'. Once this spot had been established, a kind of miniature PN junction was formed so that whisker and crystal together behaved like a diode.

Those who can remember the crystal set will certainly agree that once found, this sensitive spot on the crystal needed to be carefully preserved. No movement, no sudden shock — one almost held one's breath.

Later, the point contact diode was developed, where the cat's whisker, as it were, was permanently encapsulated in a glass tube, and today we have the germanium diode.

Fig. 53 shows this very simple form of receiver, which would be much the same as the early crystal set. It has no battery since it derives its power from the signal picked up. This of course means that only faint sounds would be heard, and then only from strong local stations — as was the case with early crystal sets. In Fig. 53 the signal frequency is selected by adjusting C, a variable capacitor, that, together with L,

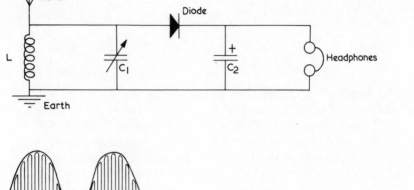

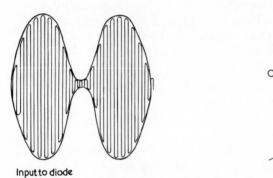

Input to diode

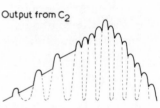

Output from C$_2$

Fig. 53. Simple diode receiver showing the detector action below. The output from C2, due to the correct choice of capacitor value and discharge time, can follow the increase and decrease of the modulating waveform

forms a parallel resonant circuit. Alternating currents at this frequency fed to the tuned circuit from the aerial will develop a magnified voltage across the high impedance. The impedance will depend upon the Q of the circuit, other frequencies fed in from the aerial will be by-passed to earth, through the inductance if they are lower than resonance, and by the capacitor if they are higher than the resonance.

In this manner, the simple receiver selects the required frequency from the unwanted one. This facet of a receiver's performance is termed its selectivity, and is a very important characteristic of modern equipment. Early receivers of this nature were dealing with AM transmission such as that from 2LO, the old London station, which existed prior to the establishment of the B.B.C.

Referring back for a moment to the amplitude modulation envelope in the last chapter, it will be noted that the rise and fall of the RF waveform is symmetrical. If an audio signal were derived from this as it stands, it would be self cancelling, since the lower wave form is in anti-phase to the upper one. The output would be nil.

The function of the diode is to rectify this signal and remove either the positive-going or negative-going half cycles. From the point of view of audio signal recovery it does not matter very much which is removed. In the receiver illustrated, the negative half cycle is being eliminated, as the positive excursions of the RF signal tuned by L and Cl make the left hand side of the diode positive, it becomes forward biased and conducts. Capacitor C2 charges up positively to a potential proportional to the peak of the input signal at that instant. During the negative excursions of the RF signal, the diode is reverse biased and does not conduct. In this period the capacitor C2 starts to discharge through the path of the headphones. The rate at which it does so is dependent upon the value of the capacitance and the resistance of the headphones. It is important that this time constant is right: neither too long nor too short. It is related to the frequency or period between cycles of the detected RF signal, in such a way that the voltage on the positive side of C2 can follow the changing peak level of the rectified signal and reproduce the audio waveform as shown in Fig. 53. The diode, the capacitor C1 and the resistance of the headphones together rectify the signal, produce the AF waveform and filter out the RF component.

The thermionic valve, a device for amplifying signals, appeared at the beginning of the century, preceding transistors by some fifty years. It made modern communication, as we know it today, a possibility. To increase the efficiency of the simple receiver, two stages of amplification can be added. One of these will amplify the RF signal coming from the aerial, and the other will amplify the AF signal coming from the detector. Such a receiver is shown in block diagram form in Fig. 54a and is known as a Tuned Radio Frequency (TRF) receiver.

The RF amplifier will enable much weaker stations to be

D*

received. The diode, because it has greater forward bias from the amplified signal, will be more efficient. The AF amplifier will enable sufficient power to be made available to drive a loudspeaker.

Although inadequate for serious operation under present-day conditions, the TRF is the type of receiver on which many listeners 'cut their teeth' and its relative simplicity made it very easy to build at home. Many of today's experienced radio amateurs, heard their first 'DX' on such simple 'home brewed' equipment.

Fig. 54a shows a basic three-block TRF receiver, to which has been added a block labled BFO which feeds into the detector. The initials BFO stand for Beat Frequency Oscillator, this unit being necessary for the reception of CW (Morse code). This is a conventional oscillator the frequency of which is adjusted to near that of the received signal, so that when the two are mixed together, the difference between them (f1 − f2) is an audible tone or beat note.

Fig. 54b shows the basic circuit diagram corresponding to the block diagram of the TRF receiver.

The aerial is coupled to the tuned circuit inductively, this being normal practice. Direct coupling would dampen the tuned circuit and adversely affect its performance. The resulting signal is amplified by TR1 and its output is inductively coupled to the detector stage. Both of these circuits are tuned for greater efficiency, and are at the same frequency. In order that both circuits will tune in step, or 'track', the two capacitors C1 and C2 are assembled on the same shaft. Capacitors assembled in this manner are described as having been 'ganged', since being mounted on a common shaft they will rotate together. Ganging is common practice in many receiver designs, and is indicated in the diagram by the dotted line.

D1 and C3 and VR1 form the detector circuit, working as previously described, VR1 being the resistance that was offered by the headphones in the first example. VR1 also acts as a variable potential divider, potentiometer, or pot for short, and determines the amount of AF fed to the last stage, which is the AF amplifier. This amplifies the signal in the usual manner, and the collector current variations drive the

106

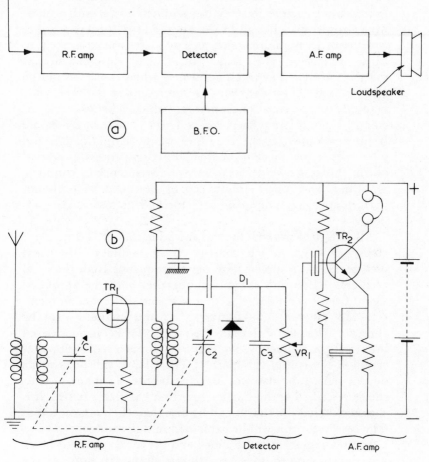

Fig. 54. TRF—Tuned radio frequency receiver
a. Block diagram
b. Circuit diagram with BFO omitted

headphone coils.

The main disadvantages of the TRF principle are:

1. Poor selectivity.
2. Limited gain.
3. Detector inefficient over the wide frequency range required of it.

Many tuned circuits and stages would be needed in order to improve this situation, but the problem is that with a number of tuned circuits, all of the same RF frequency, stability is difficult to maintain, and unwanted feedback is likely to cause oscillation. The ganging of all these different circuits together so that they track accurately when tuning would be complicated. These problems are overcome in the supersonic heterodyne receiver, or superhet, which changes the incoming signal frequency to a lower fixed intermediate frequency known as the IF. Much of the amplification and selectivity, together with the demodulation process, is carried out in this stage. As there is only one frequency to consider, amplifiers and tuned circuits can be pretuned for optimum performance, including the detector and its associated components.

All this is achieved by the introduction of a frequency changing stage, which consists of an oscillator (the local oscillator) and a mixer. The signal frequency from the local oscillator is fed into the mixer, together with the amplified signal from the RF stage, so that sum and difference frequencies will appear. The frequency of the oscillator can be chosen so that the difference frequency (f1–f2) is the desired intermediate frequency. *See* Fig. 55. This is a block representation of the complete superhet. The BFO is still present, its output fed to the detector stage as before. In many circuits this is described as the Carrier Insertion Oscillator (CIO). It is shown switched, so that it can be disabled when not required. The reason for this will be explained in due course.

In the diagram, the RF block and the local oscillator block are shown ganged together, this is necessary since as the receiver is tuned across its range f1 less f2 must remain constant, and produce a fixed IF, f2 therefore must change frequency in step with f1, and this is best achieved by ganging.

Note that the line marked AGC denotes a path going in the opposite direction to the signal. This is a form of negative feedback known as Automatic Gain Control. Its purpose is to control the gain of the receiver so that the output is constant. In practice this is not always achieved perfectly since some signals vary in strength over a very wide range. The

108

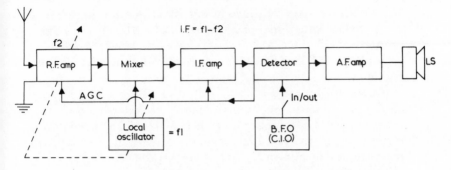

I.F = f1−f2

f2

R.F amp → Mixer → I.F amp → Detector → A.F amp → LS

A G C

Local oscillator = f1

In/out

B.F.O (C.I.O)

Fig. 55. Superheterodyne receiver (superhet) block diagram

basis of it is that a voltage is derived from the output end of the amplifying chain, usually the demodulator. This voltage is then employed to control the bias of earlier amplifying stages in such a way as to reduce the gain of strong signals and increase that of the weaker ones.

The local oscillator is shown as being higher in frequency than the RF signal, which is the usual practice. It can of course be equally made to oscillate at a lower frequency than the RF signal by the appropriate amount to produce an IF.

The value of the IF can be any frequency at all, but in practice several factors determine the frequency used, one of which is the need for a certain degree of standardisation. A typical value for broadcast AM receivers is 460 kHz, and for broadcast FM receivers 10·7MHz. Communication receivers present something of a problem in the choice of IF because of image interference and adjacent channel interference.

Image Interference or second channel interference. Image interference arises only in superhet receivers and is an inherent weakness in the principle. The IF signal is produced by mixing f1 and f2, this results in four frequencies being present at the output of the mixer, the two original frequencies, plus the sum and the difference of them, the required frequency being selected by means of a tuned circuit. As an example, if the IF is 500kHz (0·5MHz) and the receiver is tuned to a signal of 2MHz: in order to produce the desired IF the local oscillator frequency (if on the high side) must be 2·5MHz. Thus f1−f2 = 2·5MHz − 2MHz = 0·5MHz (Fig.

56a.) With the receiver tuned as described, should there happen to be a strong signal at a frequency of 3MHz this will also produce an IF since $3\cdot0$Mhz $-$ $2\cdot5$Mhz $=$ $0\cdot5$MHz. In this example, $3\cdot0$MHz is the image or second channel signal. *See* Fig. 56b.

It does not follow that there will always be a strong signal on the image frequency, but the bands are crowded and the probability is quite high. Once such a signal has been converted to the intermediate frequency there is no way of removing it, since from a point of view of the receiver it is indistinguishable from the required signal. Image interference must be prevented from getting as far as the mixer if the problem is to be avoided. Fig. 56c shows the relationship between signal frequency and the image frequency. The greater this difference, the more effectively will it be attenuated by the tuning circuits in the RF stages. Since D is twice the IF then the higher it is the better the image rejection.

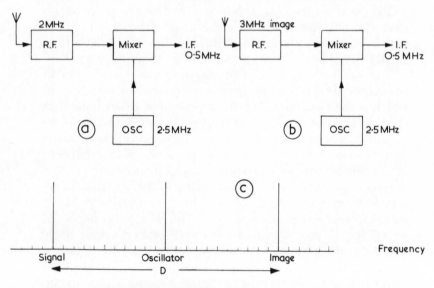

Fig. 56.   Image or second channel interference
      a.   Correct relationship
      b.   Image relationship
      c.   Shows difference between signal and image. The greater D becomes, the better is the image rejection

Adjacent Channel Interference. Unlike image interference, adjacent channel interference is concerned with selecting the tuned frequency and attenuating strong signals immediately adjacent to the one required. This can be a very exacting requirement for a communication receiver, perhaps the most important of all. Much thought and effort have gone into the design of filters and selectivity in modern equipment. Ideally the response of the receiver would be just wide enough to cover the band of frequencies required for intelligible AF, and no more, this amounts to a characteristic response with vertical sides (*see* Fig. 57). This steep sided shape is not achievable with inductive capacitive tuned circuits. However, since practically all the selective circuits are in the IF stages, much can be done to approach this ideal. The bandwidth of a tuned circuit decreases as the frequency of resonance decreases, therefore a low IF will improve selectivity.

This then is the problem, high IF is desirable for good image rejection, and a low IF for good selectivity characteristics.

The solution with communication receivers is either a compromise frequency of perhaps 1·6MHz or the use of the double-superhet. Here the first mixer produces a high IF for image rejection and the second mixer converts this frequency

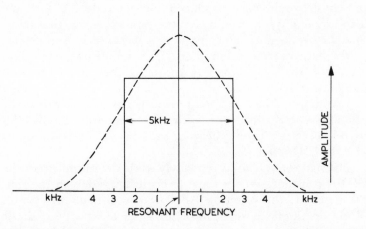

Fig. 57. Ideal bandpass response in full line. A more typical response in broken line

to a very low IF where the selective tuned circuits are situated.

Much use is made of crystal filters and mechanical filters in modern equipment. Crystal filters are sets of crystals specially matched and factory-set to give a steep sided response and narrow audio bandwidth. In the mechanical filter, electrical energy is converted to mechanical energy in a series of discs, then converted back to electrical energy, again factory-set and capable of producing steep sided, flat-top band pass characteristics.

After the signals have passed through the frequency changing stage and have been selected and amplified in the IF stage they are demodulated, and the resulting AF is amplified. The AF amplifier will be a conventional stage, its output driving a loudspeaker or headphones. The demodulation for amplitude modulated signals will be basically the same as the detector described earlier; however, to resolve SSB signals an extra process is required. In SSB the carrier and one of the side bands has been suppressed at the transmitter stage. Although the carrier contains no audio information in itself, its presence is vital to demodulation, since the audio component is the mathematical difference between the carrier and the upper or lower sideband. It has therefore to be restored within the receiver. Had the carrier been present it would have produced, by heterodyning with the local oscillator, a frequency corresponding to the intermediate frequency. This frequency, because it is constant, is where the carrier is reinserted. This is done by means of an oscillator which is commonly known as the carrier insertion oscillator (CIO).

The simple AM detector previously described is not really efficient enough to cope with the demodulation of SSB mainly because of the difference in levels of the signal inputs and the CIO input.

The signal levels will vary widely while the output from the CIO is constant. Where these levels differ widely in the simple detector, the resulting AF signal is difficult to resolve. Many operators however have used the older type of basic communication receiver not designed for SSB and by skilful use of the controls and through familiarity with the equipment have for years copied SSB signals quite satisfactorily.

The modern communication receiver may employ a product detector for the resolution of SSB and CW and a diode detector for AM. It should be noted that SSB signals are easier to tune and less distorted with this refinement. The product detector provides a mixing process where the two signals to be mixed are the sideband signal coming from the IF amplifier and the output of the CIO. The CIO frequency is set so that the frequency between these two signals is the AF required. An example of the basic product detector is shown at Fig. 58. C3 the $0 \cdot 5 \mu F$ capacitor is needed to couple the AF output of the detector and block the FET drain DC voltage to the next stage. C1, C2 and the RFC filter out the unwanted RF signal. The initials RFC stand for Radio Frequency Choke, an inductance which has a high reactance at radio frequencies.

Finally, there is frequency modulation which is basically different from AM in that the amplitude is held constant and the frequency allowed to vary or deviate at an AF rate. This requires a different type of detection, hence the principle differs from those previously considered, which are unsuitable for this purpose. Many receivers in the 1–30MHz range have no FM detector fitted. For years this was not a popular form

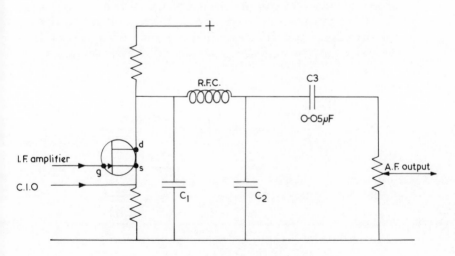

Fig. 58. Product detector using FET

of modulation among radio amateurs, but with the increasing popularity recently of VHF bands NBFM has become more common. It is possible to resolve FM on a receiver in the AM mode by slightly detuning to one side of the carrier frequency. This is known as slope detection and occurs by an accident of band pass characteristics. The result can be poor because the receiver is operating at a lower gain point and there may be distortion because the detection slope is not linear.

The most efficient demodulator of FM signals is the discriminator. Many circuits have been designed and basically they consist of a limiting stage, the purpose of which is to remove any amplitude modulation which may be present. Much of the random noise and interference is amplitude modulated and limiting can remove most of this, making FM a clean signal. Following the limiter is a circuit for converting frequency changes into amplitude variations. One basic circuit for doing this, and there are many variations, is shown in Fig. 59.

Essentially this depends on circuit A and circuit B which are tuned to the exact carrier frequency. With the advantage of the superhet receiver, this frequency will be the IF, the input from which appears across the circuit A. This is coupled to its secondary circuit B. At resonance, which is the frequency of the unmodulated carrier, this circuit is resistive. Half its voltage appears across diode D1 and resistor R1, while the other half appears across D2 and R2. The voltages are

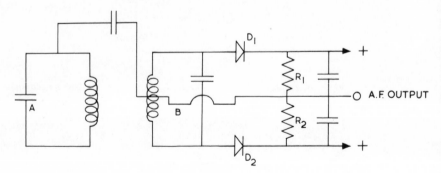

Fig. 59. A basic arrangement of an FM discriminator

equal in amplitude and opposite in phase. The rectified voltages across are R1 and R2 and are therefore equal and opposite, and the output is zero. With modulation the carrier frequency deviates and the tuned circuit changes from resistive resonance to reactance. As deviation increases the carrier frequency circuit B becomes inductively reactive, as it decreases the carrier frequency it becomes capacitively reactive. The voltages across D1 and D2 differ in proportion to this change in phase. The voltages across R1 and R2 are no longer equal and the difference between them is an audio frequency output, whose frequency is equal to that of the frequency of deviation.

Chapter Ten

RECEIVERS—Part II

When examining modern communication receivers, their specifications and circuit diagrams, you will find many complicated working variations from the simple principles we have examined here. Much of this sophisticated circuitry is concerned with making the receiver more efficient in coping with today's crowded air waves and exacting requirements. Let us therefore look briefly at some of these in more detail.

Band Spread. This is an aid to easier tuning of signals which may be very close together, essential for the resolution of SSB signals. Remember that when tuning such a signal the accurate positioning of the sideband in relation to the reinserted carrier is crucial, since the difference between them is the audio intelligence. Any error results in distortion—the familiar 'Donald Duck' sound. In tuning across a frequency band of say 14 to 14·350 MHz, the exact selection of a frequency to within a few hertz calls for a very high quality tuning mechanism with no slack or backlash. To assist such tuning, band spread may take two forms—mechanical or electrical. Mechanical tuning usually depends on low gearing between the control knob and the component being varied. As previously stated, the tuning of the receiver consists of making the resonance of one or more parallel tuned circuits change. This is done by varying the inductance or the capacitance, and if necessary ganging several variable components together. Coupled to this mechanical linkage is an indicator on the receiver front panel which shows how the frequency is changing. It may take the form of a horizontal scale (Fig. 60a), across which moves a vertical line called the cursor. Or as in (b), the cursor is fixed and the circular scale rotates behind it. Where low gearing provides the band spread by allowing the cursor to move only a small amount with several rotations of the tuning knob, a problem arises when it is required to move quickly from one end of the band to the other. This is sometimes overcome by providing a handle at right angles to the tuning knob in order that it may be rapidly rotated. Another system is to have a very free running tuning mechanism and to provide a weighted flywheel within the set. When a sharp twist is imparted to such a tuning arrangement, the flywheel will spin and its momentum will carry the tuning from end to end. A further method is to provide a two-gear system, one providing coarse tuning and the other fine. The selection is usually achieved by means of pushing in or pulling out the main tuning knob.

Electrical bandspread is achieved by providing a variable capacitor of very low value in parallel with the tuning circuit(s), such a capacitor, being in parallel, will add to the value of the main capacitor, but only by a small percentage. This component will be brought out to the front panel and controlled with a separate knob, called variously bandspread, fine tune, clarifier; they all mean and do the same thing (Fig. 60b).

Many receivers will have combinations of any or all of these various systems of bandspread.

One final method of bandspread in addition to the foregoing is the amateur band only receiver, this is fine for licensed operators and listeners who confine their interest to the amateur bands, since the whole tuning range can be arranged to contain just these frequencies. This, together with the other methods described will give a remarkable degree of precise tuning. For many people, however, the flexibility of a general coverage receiver, that is one which receives all frequencies within a given range, will appeal more than the limited amateur band version. The modern communications receiver is an expensive item, therefore much consideration must be exercised as to what is expected of it.

Calibration. This is an essential facility in any amateur or shortwave listener's receiver. It is the means by which the cursor setting on the scale is compared with a reference signal to check for accuracy. In many stations the receiver is the standard by which the transmitter is compared. Inside the set there is a crystal oscillator designed to have a very high degree of accuracy and stability. If the fundamental

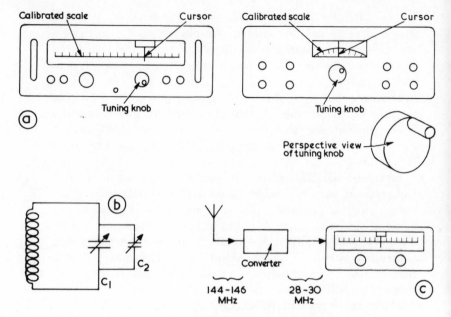

Fig. 60. a. Typical receiver control layouts
b. Electrical bandspread or fine tuning
c. Converter, 2 metres to 10 metres

frequency is 100kHz and the oscillatory circuit is of a type rich in harmonics, useful signals can be produced up to 30MHz or so. This means that when the calibrator is activated by means of a push button or switch on the front panel, and the BFO or CIO switched on, beat notes can be heard every 100kHz along the band being tuned. In addition to this facility there must be a means of adjusting any small errors that may occur. The usual method is to provide a small amount of adjustment for either the scale or the cursor, independent of the tuning circuit itself, so that if at zero beat with one of the calibrator harmonics the cursor does not line up, it can be moved relative to the scale until it does.

Digital Display. The foregoing are called analogue displays but becoming ever more popular is the digital display. Here the frequency being tuned is converted into

118

signals which drive an LED (Light Emitting Diode) display. This is of course far easier to read at a glance, is difficult to misinterpret and gives an accurate readout to 100Hz or so. Many receivers incorporate both systems.

Variable Selectivity. For AM reception, a wider bandwidth is necessary than is required for SSB, where only one sideband is being received. For the former the requirement is approximately 6kHz, while for the latter 2−2·5kHz is sufficient. For CW a very narrow bandwidth of 100Hz or so will still enable the signal to be copied. The narrower the bandwidth that can be accommodated, the more selective the receiver will be. At the same time the bandwidth has to match the type of emission in use. AM would be impossible to copy with a narrow bandpass characteristic, so many receivers have three or more positions of selectivity. Often, as mechanical or crystal network filters are expensive, they may not be fitted but available as optional extras.

Stability. Is the measure of the receiver's ability to stay on the frequency to which it is set. It is most important when copying SSB signals, since small amounts of drift will cause audio distortion as mentioned earlier. Also when receiving CW in a narrow selectivity mode, any drift will cause the signal to disappear outside the passband altogether. Many circuits and techniques have been devised to keep frequency drift to a minimum. A figure is usually quoted as being less than a given number of hertz in an hour, after allowing a period of warm-up.

RF and IF Gain. In addition to the AF gain or volume control, RF and sometimes IF gain are found on communication receivers. These provide a much greater degree of control than would be possible by just leaving these gains to the AGC system. Very strong signals (especially SSB) are easier to resolve if the RF gain can be independently set.

Signal Strength Meter. This is a small dial calibrated in 'S' points with a scale indicating the relative strength of the received signal. It is useful for making comparisons between transmissions from the same station, under different conditions of weather, time of day, aerial, etc. There is some argument about the value of 'S' meters, and it is true that their indications should be interpreted with a certain

119

amount of caution. S units are the numbers in a sequence from 1 to 9 which are used to report the strength of a received signal. S1 means faint signals, barely perceptible—ranging through to S9, which means extremely strong signals.

Converter. The name given to a unit, sometimes built in, but usually separate, which in essence contains a frequency changer. It enables signals beyond the frequency bands covered by the receiver to be tuned. It is placed between aerial and receiver and by a frequency changing process converts signal frequencies up or down to a band that is covered by the receiver. The oscillator in the converter has a fixed frequency, in fact it is crystal controlled. An example is the two metre converter (Fig. 60c). Signals in the band 144 MHz to 146MHz are fed into the converter where they are heterodyned and their output, in this example, will be within the band 28—30MHz. This band is then selected on the receiver and tuned, when it will receive the two metre signals. The lead between the converter and the receiver should be short and screened to prevent the pickup of external signals in the 28—30MHz band.

AGC Amplified. In order to achieve a greater range of control the voltage that is derived from the last IF stage is amplified before being applied to the earlier stages of gain.

AGC Delayed. In order to prevent AGC action from affecting the very weak signals, the diode producing the control voltage is reverse biased to a degree, such that until the signal arriving from the IF is large enough to overcome the bias no AGC is produced.

AGC Fast and Slow. With AM signals there is a constant carrier producing a proportional control voltage. With CW or SSB signals, however, the signal comes in short bursts of RF energy. AGC systems designed for these modes ideally have a fast attack time, but a slow decay time so that the control voltage has not altered the gain between the periods of interrupted signal. A typical specification for AGC characteristic is attack time 1ms. Release time variable 100mS—1 sec.

Noise Limiter. Much of the man-made and natural noise occurring on the amateur wave bands is amplitude modulated, consisting of spikes exceeding the average height of the envelope wave form. The limiter works by passing the signal through a circuit which is able to cut off or limit these positive and negative peaks. *See. Fig. 61.* Of course the limiting must not have too great an effect on the amplitude modulation, or severe distortion will occur. Many very sophisticated noise-reducing circuits have evolved, and the more comprehensive communication receivers will almost certainly have such circuitry included.

The Headphone Output. This is always included. Many listeners find headphones essential for the copying of weak signals, especially CW, since it aids concentration by excluding surrounding room noise. Some communication receivers have no self-contained speaker and therefore an external speaker output is also included.

Recorder Output. This is also often found, since it is of great interest to many listeners to be able to record directly from the receiver signals for analysis or perhaps for replaying to the originator of the signal. Remember an amateur has no way of knowing what his own signal sounds like at the receiving end, unless he is able to listen to a recording of it.

12V and 240V. Since modern equipment is virtually all transistorised it is convenient to have the supply designed for 12 volts, thus making it possible to operate from car battery sources in caravan, boat, tent, etc. There is usually a built-in 240V conversion circuit as well, although this is sometimes

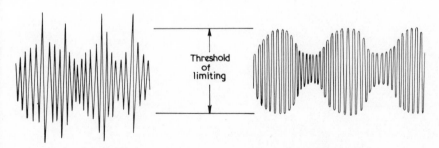

Fig. 61. Modulated RF signal with noise spikes before and after noise limiter circuit

an extra outboard unit, for operation from the domestic mains supply.

Muting. This is a facility for rendering the receiver less sensitive for periods of transmission without turning all the circuits off. For stability it is preferable to keep oscillator circuits operating. Muting usually works by having two contacts at the rear of the receiver opened and closed by the operation of a relay controlled from the transmitter. When opened, a large bias is introduced into one or more of the RF amplifying stages of the receiver, thus greatly reducing its sensitivity. The amount of this muting is usually adjustable. When the contacts are closed the muting circuits are by-passed and the receiver resumes normal gain.

These are some of the more common facilities to be found in modern communication receivers. A glance at any manufacturer's specification sheet will often show many more, and as technology develops, others will be added. Often they are little more than gimmicks, or old ideas dressed up in new trimmings.

When considering the purchase of new equipment, try to determine whether or not the features offered will positively assist in the better reception of weak signals in the presence of interference. This is ultimately what is most important.

Chapter Eleven

TRANSMITTERS

The transmitter is a device for generating radio frequency signals capable of carrying intelligence. This intelligence may take the form of audio modulation (as with amplitude modulation) or frequency modulation. It may simply be keyed on and off to form a burst of radiation which can be sent in the form of Morse code.

There are other forms of modulation and intelligence which can be transmitted, some of them available to the amateur under the terms of his licence. We have concerned ourselves with audio information and Morse code, since these are the ones mostly found on the amateur and short wave bands.

Another type of emission is RTTY (Radio Teletype), which sounds like a musical high speed Morse code and is achieved by frequency shift keying (FSK). That is to say the bits of information radiated are defined by a change of frequency, rather than on-off keying as in CW. In place of the microphone in such a transmitter is a special typewriter keyboard, and when a key is depressed the coded form for that character is radiated. In the past, the receiver of such a signal was coupled to a device which recreated the character on paper, rather as a typewriter does, or in continuous strips. Nowadays the signal is converted into information which can be fed into a video display unit, rather like a television. Video transmissions are also allowed in the amateur bands, and although of low definition compared to the broadcast systems, they nevertheless allow amateurs to relay pictures to one another. Radio transmissions can also be modulated, so that control information can be conveyed. Radio controlled model aeroplanes and boats are a simple example. More complex is the control of space craft and the recovery of information from other planets, all of which has been done in recent years.

Navigational aids such as RADAR, automatic landing

systems, location or altitude indicators, are all ways in which RF transmission is used.

Returning to the basic AM transmitter, Fig. 62a, we see the three basic blocks representing the oscillator, the multiplier or buffer amplifier, and the power amplifier. Also indicated is the audio amplifier modulator, and this feeds into the PA stage. The Morse code key is shown controlling the buffer stage. This diagram shows the usual arrangement of an AM transmitter.

The oscillator circuits are operated at a low-power level and a low frequency to ensure stability and accuracy in operation. Usually, the oscillator is variable, allowing the transmitter frequency to be adjusted to that of an incoming signal. This is known as netting. Frequency stability is essential to prevent the transmitted signal from drifting outside the permitted band of operation. It is important, too, to have a signal which can be easily resolved by the listening station. A signal which drifts outside the pass band of the receiver tuned to it will be lost.

This can happen when two stations are netted to one another to begin with. Then if one of the transmitters is drifting during a period when it is on standby, when the other station passes the transmission back, the offending transmitter has drifted away from the frequency to which the listener is tuned. Communication under these conditions is difficult. Also of course, the signal may drift from what was a clear channel into an occupied one.

So frequency stability is very important. Factors which can cause the oscillator to drift in frequency are lack of rigidity in the components forming the circuit, especially the capacitor and inductors, and poor mechanical construction. Any mechanical movement due to shock or vibration will change the value slightly, and hence the frequency of resonance. A VFO (variable frequency oscillator) should have a soundly constructed variable capacitor or inductor, especially important in mobile equipment where vibration is an obvious hazard.

Temperature change is another cause of drift, since with a change in temperature there is an accompanying expansion or contraction of materials from which the inductor and capacitor are made, and again these changes will result in a shift of the resonant frequency. The effects of heat may be minimised by keeping the equipment well ventilated, by running the oscillator stage at low power so that little heat is produced by the semiconductor, and by keeping the transmitter away from sources of external heat, such as a radiator or table lamp bulb. The location of mobile equipment in cars is again a potential problem, and care must be taken to allow for ventilation.

Voltage variation. Any change in the voltage supply to the oscillator transistor will cause a small change in the characteristics of the device. For example, the capacitance will change slightly and as this is associated with the tuned circuit the resonant frequency will change. A stabilised voltage supply to the oscillator stage is the usual practice. The stabilising is achieved by means of a zener diode in many circuits. You will come across stabilised power supplies which are much more complex than the simple zener described earlier. The reason for this is that a power supply has to stabilise voltages over a wide range of current variations, and this the basic zener circuit could not do. However, as the oscillator is deliberately run at very low power, the current changes associated with it will be negligible, and in this application the zener is suitable.

The oscillator in a transmitter may be crystal controlled, thereby taking advantage of the stable characteristics of the quartz.

The disadvantage of course is that the frequency is fixed, and to achieve a range of frequencies one would need a number of crystals connected into a circuit via a rotary switch. Netting is then not possible unless by chance one of the crystal frequencies happens to match. Nevertheless, many transmitters, especially those in the VHF bands and many aircraft and maritime sets, are crystal controlled.

The next block in Fig. 62a is the buffer amplifier or amplifier. In a simple transmitter this stage, or stages, will act to amplify the output from the oscillator (which is of a low output) to the level necessary to drive the power amplifier. It is also desirable in the interests of stability to separate the PA from direct connection with the oscillating circuits. The

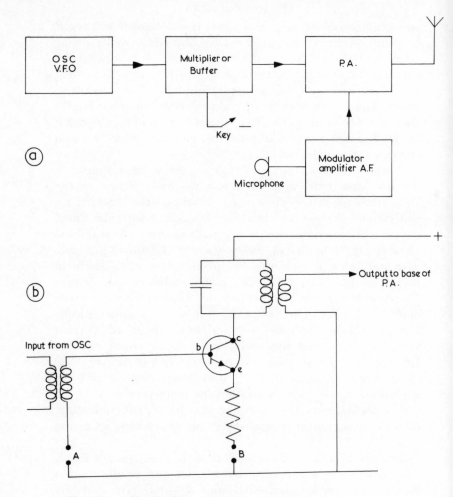

Fig. 62. a. Simple AM transmitter
b. Buffer/amplifier (class B) showing possible keying points at A or B

buffer stage accomplishes this too.

In a CW (Morse) transmitter this is also the ideal stage at which to key the transmitter. By doing this the oscillator is left running during the key-up periods. Keying the oscillator on and off is likely to produce a characteristic known as 'chirp', which is a very apt description. This bird-like noise

126

is produced because the oscillator has to start up each time the transmitter is keyed and what you hear is the frequency changing as it does so. The power amplifier will be passing a high current, especially in transmitters operating near the maximum of 150 watts, and keying high current is undesirable if it can be avoided. Fig. 62b shows a buffer amplifier, operating in Class B, and with two possible keying points indicated.

There may also be one or more stages of frequency multiplication following the oscillator, and it is convenient that the LF amateur bands are harmonically related. Therefore, if the basic oscillator frequency band covered is $1 \cdot 75$ MHz to $2 \cdot 00$ MHz, which embraces all of the 160 metre band, all the other LF amateur bands can be achieved by multiplying. Observe that some of the stages cover a band of frequencies wider than that allocated to amateurs. With such a system, care has to be exercised to ensure operation within the band.

Multiplier Basic	Band produced MHz	Amateur Allocation	Name
	$1 \cdot 75 - 2 \cdot 00$	$1 \cdot 80 - 2 \cdot 00$	160 metres
×2	$3 \cdot 50 - 4 \cdot 00$	$3 \cdot 50 - 3 \cdot 80$	80 metres
×2	$7 \cdot 00 - 8 \cdot 00$	$7 \cdot 00 - 7 \cdot 10$	40 metres
×2	$14 \cdot 00 - 16 \cdot 00$	$14 \cdot 00 - 14 \cdot 35$	20 metres
×3	$21 \cdot 00 - 24 \cdot 00$	$21 \cdot 00 - 21 \cdot 45$	15 metres
×2	$28 \cdot 00 - 32 \cdot 00$	$28 \cdot 00 - 29 \cdot 70$	10 metres

The final stage in the AM transmitter is the Power Amplifier. At present most of the transmitters in use have valve amplifiers if the PA is of significant power, say 100 watts or more. No doubt future developments with semiconductor devices will bring more powerful transistors into the scope of amateur equipment. For some time to come the operator is likely to find either of these devices in the final PA.

The characteristics of valves and transistors are somewhat different, especially at high powers, and the output circuits therefore vary. The DC power input to the aerial coupling circuit, as limited by the licence conditions, is measured at the output of the final amplifying device. This means that the power determined by the product of voltage and current at this point should not exceed that specified in Appendix B to the licence. This means in effect 10 watts for Top Band and 150 watts for the other bands in the above table. Valves, generally speaking, operate with much higher voltages than transistors, and therefore at any given power the transistor PA will carry a higher current than its valve counterpart. The input and output impedances of valves are generally high and with transistors low in PA stages.

In either case the load of the PA is a resonant tuned circuit whose capacitor or inductor or both are variable to bring the circuit to resonance. This point is indicated by the drop in current displayed by a meter in the voltage supply to the amplifier. When a parallel tuned circuit is resonant, it becomes high impedance, and there will therefore be a drop in the current flow. Amateurs refer to establishing this point as 'tuning for a dip'. Many modern transmitters have broad banded output circuits, with a constant output, requiring no adjustment since such a PA needs no tuning.

The output of the aerial terminal may be capacitively or inductively coupled, but is invariably arranged to have an impedance of 50–75 ohms.

High level modulation is achieved by passing the audio signals generated by a microphone through an audio amplifier, restricting the bandwidth to about 3·5kHz, and coupling the output to the PA stage by means of a 1:1 modulation transformer.

The disadvantage of amplitude modulation is that, apart from the wide bandwidth it produces, the size and power requirements of a modulator are much the same as that of the transmitter itself.

Frequency Modulation. On the VHF bands this system has become very popular: it has two distinct advantages. It is very simple to carry out the modulation process, just a few extra components are needed and for this reason negligible weight is added to the equipment. This obviously makes it advantageous for portable or mobile use, and the VHF bands make it doubly so since the aerials required are very short. We shall see the importance of the aerial feature in a later chapter.

1. *Above* QSL 'Wallpaper'. Cards from around the world often decorate the walls of amateur shacks.

   Below Inside view of QRP (low Power) CW only transceiver, built from one of the excellent kits by Heathkit. This is the HW8; it will work the world on less than three watts!

2. *Above* Rear view of home-built equipment. SWR bridge on left, simple antenna tuning unit on right.

Below Home built SWR bridge (left) and amateur bands receiver (right), by G3UWJ. Note dummy load bottom left of rig.

3. Compare the picture in 2 *below* with this view. Here the chassis has been removed from the cabinet to show the construction of the receiver. Note the valves.

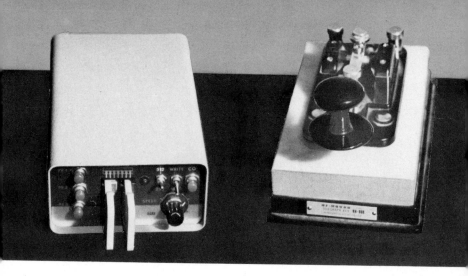

4. *Above* Great pleasure can be obtained from CW operating. Traditional Morse Key on right. Automatic key on left, with automatic 'memories'.

Below A very useful piece of equipment for the amateur shack. A dual trace oscilloscope, by Trio.

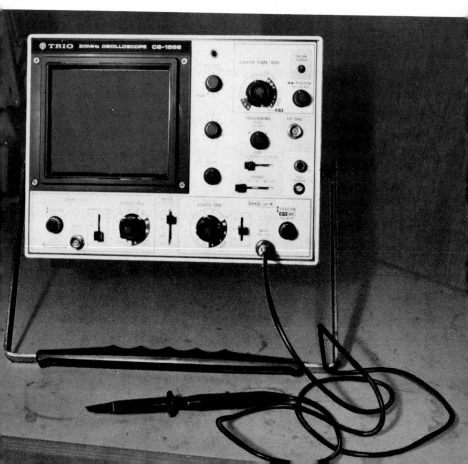

5. A good antenna tuner is a big help. Scale on left shows SWR (Standing Wave Ratio). Scale on right shows forward power in watts. Adjustment is made for minimum SWR and maximum power output.

6 & 7. Amateur stations can be complex and expensive, as here — or may consist of just one home-built rig. Tremendous pleasure can be derived from either.

8. Sophisticated SWR and Power meter; needles indicate reflected power and forward power. SWR can be read off at their point of intersection. Note also the Frequency Counter (YC 355D by Yaesu) which covers 30 to 200 MHz. The frequency is displayed digitally.

FM signals are usually generated at a lower frequency than the radiated one, and then multiplied in the usual way. A typical 144MHz transmitter is shown in block diagram form at Fig. 63a. The audio signals produced by the microphone are amplified and applied across a variable capacitance diode. The diode changes its capacity in proportion to the varying RF signal. The varicap is connected to the frequency determining components, usually a crystal, and this will cause the frequency to deviate by a small amount. If the oscillator is an 8MHz one, then it will have to be multiplied 18 times to achieve the output of 144MHz. This means the deviation is also multiplied 18 times. Therefore a deviation of only 139Hz when multiplied produces a shift of $2 \cdot 5$kHz, that recommended for NBFM (narrow band frequency modulation).

Low power outputs up to 30 watts or so will usually be transistorised, and the input network designed to match $50\,\Omega$.

The SSB transmitter is probably the most widely used in amateur radio. Fig. 63b shows the stages in block diagram form. Basically it has to do two things, eliminate one of the sidebands and suppress the carrier. There are different ways of achieving this, but by far the most common way in current equipment is the Filter Method.

Referring to the diagram, you will see that the first stage is the basic oscillator which may be fixed or more probably a VFO running at the lowest frequency practicable as before. The microphone at the left is shown connected to an AF amplifier and the modulation takes place at this early low power level in the transmitter. In this case there is no high power bulky modulation stage. The mixing process produces the sum and the difference frequencies $f1 + f2$ and $f1 - f2$, i.e. the upper and lower sidebands. However, the mixing takes place in a special kind of modulator, called a balanced modulator, the effect of which is to cause the carrier frequency to be cancelled out, leaving an output of just the two sidebands. Fig. 63c shows the circuit of one such modulator. The RF input is fed in via the mutually coupled tuned circuits and applied across the bridge circuit formed by D1 and D2 and VR1. When the circuit is balanced

E

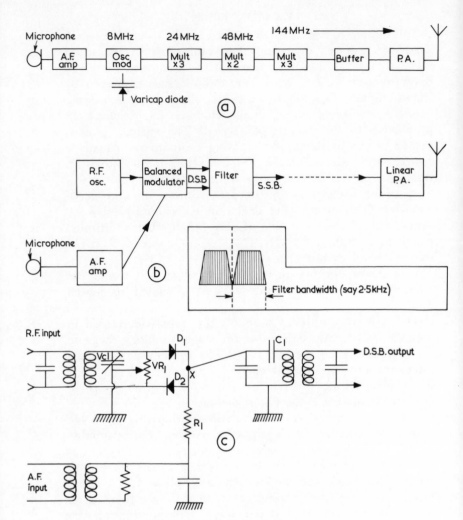

Fig. 63. a. 144 MHz FM transmitter
          b. Single sideband transmitter (filter method)
          c. Balanced modulator

there is a null at point X, the centre of the diodes, and the RF signal is cancelled out. This is the carrier suppressor. To help achieve this and to allow for any difference in the characteristics of D1 and D2 (which ideally should be identical) the centre tap of VR1 is adjustable. Some variable capacitance VC1 may also be included.

Introduction of modulation unbalances the bridge, causing the diodes to conduct alternately at an audio rate. The result is a double sideband (i.e. upper and lower) output via the mutually coupled tuned circuit. C1 is added to prevent the audio input via R1 being short circuited through the inductance.

The unwanted sideband is suppressed by passing the signal through a selective filter. In order to be effective, the filter must have a good shape factor, i.e. have steep sides and a bandwidth of about about 2·5-3kHz. Such filters are fairly complex and therefore costly, but by this method very effective sideband signals can be produced.

Since the RF signal now carries audio information, subsequent amplifying stages must be linear to avoid severe distortion. For the same reason frequency increasing is carried out by heterodyning and not by frequency multiplication. Usually, in an SSB transmitter the RF signal is generated by a crystal controlled oscillator, the wanted sideband is produced as outlined above. The signal is then hetrodyned in one or more stages to the required final transmitting frequency. One of these frequency changing stages will have a VFO in order that the transmitter may be tuned over the span of the band in use.

Technically, there is no difference between the upper and lower sidebands and either may be radiated. It is conventional however to use the lower sideband in the three lower amateur bands, top band, eighty metres and forty metres, all the higher bands using the upper sideband.

In January 1982 a new set of codes for specifying emission came into official use. The types you are most likely to come across are as follows: —

Amplitude Modulation.

A1A   Telegraphy by on-off keying.

A3E   Telephony, double sideband.

J3E   Telephony, single sideband, suppressed carrier.

Frequency Modulation.

F3E   Telephony.

A full list of the emission codes will be found in the booklet *How to become a Radio Amateur* (*see* page 11).

Thus J3E Amplitude modulated telephony single side-band suppressed carrier (the most usual form of **SSB** heard on the bands today). The transmitter is a device built for radiating RF frequencies. It is essential, indeed obligatory, on the part of the licenced operator to radiate only those frequencies and at those power levels defined in his licence. It is an offence to radiate other frequencies and to exceed the power limits, whether they cause interference or not. Within the transmitter, as we have seen, signals are generated other than those finally to be radiated.

Therefore the construction of all commercial equipment is of metal and forms a complete screen, which may be earthed, around the entire circuitry. Internally, frequency changing stages will all be screened from each other and from the outside. Leads such as those for the microphone and Morse key should also be screened, and also filter circuits at the point where all external leads leave the equipment to prevent the escape of spurious signals by these routes.

Most current equipment has a very good record for its interference-free operation, but as we shall see later, there are other ways in which interference can be created. Those who embark on their own construction of transmitters need to pay very special attention to these matters and be guided by the advice given in the RSGB handbook.

Chapter Twelve

TRANSCEIVERS

A visit to an amateur's 'shack' — the colloquial name by which the operating room or area is known — or a reader of the amateur press, will soon realise that a great deal of operation takes place with the 'transceiver'. This is simply a box which contains both the transmitting and receiving functions.

Strictly speaking, there are two distinct types: the transmitter-receiver, which is in effect two separate items contained for convenience and economy in one case. The transceiver is a more complex arrangement where much of the circuitry, oscillators, frequency changers and amplifiers, for example, are shared between the two functions, the changeover being effected by relays and solid state switching (diodes) when the operator pushes the transmit button.

In both cases some of the controls and instruments are common to either function. For example the tuning dial or channel selector will show both the transmitted and received frequency. In some designs there is provision for the transmitter or the receiver section to operate at a controlled offset frequency. In some transceivers there is a RIT control which stands for Receiver Independent Tuning, which allows a small variation of frequency from that which is set on the main control. In two metre equipment it is usual to provide a 600kHz offset, in order that repeater working can be accomplished. This is covered in more detail later. The meter movement will show signal strength on receive and relative power output, with perhaps other information on transmit.

Why has the transceiver become so popular? It has the obvious advantage of convenience and economy of space, as already mentioned. Because many of the components perform dual functions there will be an economy of cost too. Many amateurs like to operate from a car, boat, or caravan, and the transceiver lends itself ideally to this.

Many operators too, for reasons of limited accommodation, have to share their shack with a bedroom or living room and the advantage of combining both units in one package is apparent.

There are also advantages in the separate transmitter receiver system. Greater flexibility of split frequency working, i.e. the frequency of transmitting need not be the same, or even on the same band, as the received frequency. When the units are entirely separate, there is a greater choice of specification for each component, choosing the functions considered most important in each. The two items may even be of different manufacture. Many amateurs prefer their receivers to be of the general coverage type and not limited, as the receiver section of a transceiver is, to amateur bands only. This is really a simplification, but it serves as a guide to the basic differences.

Many transceivers and transmitters can be VOX (voice operated transmission) controlled. Part of the amplified signal from the microphone is used to operate the transmit/receive circuitry. The system stays in the transmit mode for a short period after the audio signal ceases, to allow for natural pauses between words and sentences. This delay is adjustable to suit individual requirements. When the operator has finished his period of sending or 'over', his silence releases the VOX, and the equipment drops back into the receive mode. Such systems need a sensing circuit which enables the transmitter to recognise its own receiver, so that noise from the speaker does not activate the VOX. This sensitivity is adjustable.

Chapter Thirteen

ANTENNAS

The aerial or antenna is probably the most important component in the whole station. It is sometimes the most neglected, and nearly always a compromise between the practical and the ideal. The theoretical aerial is the same for receiving and transmitting, and therefore a good transmitting aerial is a good receiving one. In most amateur arrangements, for any one band, the same aerial serves both purposes, being switched from one function to the other by electronic means.

A simple aerial was mentioned in the section on propagation, where the alternating RF currents were being fed to the vertical wire. This wire then caused the magnetic and electric fields to radiate. Because there is an alternating frequency factor, the question of resonance arises, and in fact for any given radio frequency there is a resonant aerial. The length of this resonant conductor is related to the wave length of the radiated frequency. For maximum efficiency and proper transfer of energy, the aerial must be resonant, or as close to resonance as possible. This means that any odd length of wire thrown out of the window is most unlikely to be suitable for transmitting purposes. The fact that such a random length may appear to be satisfactory as a receiving aerial is due to the fact that the resonant factor is not so critical on received signals, and most receivers have good enough sensitivity to make up for the short fall in aerial performance. However, a resonant antenna will perform much better, and other benefits such as the rejection of noise and interference will accrue.

Let us therefore examine the theoretical resonant aerial. The electrical length of an aerial at which it is most efficient is a multiple of half wavelengths. This is related to the speed of propagation (300×10^6 metres per second), referred to previously. The electrical length is not necessarily the physical length of the aerial. Adding inductance in the form of a coil can make a wire electrically longer, and adding a capacitor makes it electrically shorter.

The electrical length is also modified by the end effect caused by the presence of insulators and supports which of necessity have to be in close contact with the wire.

The thickness of the wire itself can modify the bandwidth and length slightly.

The basic aerial is the half-wave dipole (*see* Fig. 64a), so called because it is split into two parts and fed at the centre. Such an aerial will have an electrical length of half a wave and is said to be resonant at this frequency.

The length of an aerial can be worked out from the formula:

$$\lambda = \frac{300 \times 10^6}{f}$$

and the result divided by two to give the half wavelength. In practice, to allow for the factors mentioned above and owing to the different velocity of radio waves in wire, compared with free space, the actual length is about 5 per cent shorter than this.

The current and voltage waveforms are shown in Fig. 64a. In a resonant antenna these give a low impedance at the centre. This is about 75 Ω in a straightforward dipole in free space away from buildings and trees. This 75 Ω is known as the radiation resistance of the aerial. There is no actual resistance of course. It is the resultant of the current which is highest at this point and the voltage which is low. Ohm's Law says $R = \frac{E}{I}$, so that at the centre where E is low and I is high, the apparent resistance is about 75 Ω and the power dissipated in such an aerial by a transmitter would be exactly the same if the aerial were replaced by a resistance of this value. Such a device is often used for testing purposes when it is called a 'dummy load'.

If such an aerial were fed at one end, then the result of $R = \frac{E}{I}$ would be high — 100,000 Ω or more. Modern transmitters all have a low impedance output, 50 Ω – 75 Ω, designed to feed their power into the low impedance point in the dipole, where the transfer of power is safer and more efficient. Feeding high impedance and therefore high

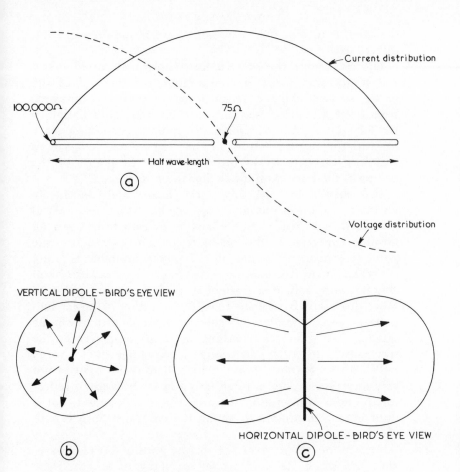

Fig. 64. Basic dipole aerial with vertical radiation pattern (a) and
horizontal radiation pattern (b)

voltage from the transmitter can lead to losses through leakage, sparking, flashover, and also the hazard of shock. If therefore, as often happens, an aerial is fed at its high impedance point, the transmitter has to be coupled to it through an impedance matching device. One such design is in fact called the 'Z' match (Z = impedance).

The electromagnetic radiation from a dipole is at right angles to its length and equal in all directions, so if the dipole is mounted vertically and the radiation pattern is

E*   137

depicted as in Fig. 64b, it is said to be omnidirectional (all directions).

Fig. 64c shows the pattern for a horizontal dipole. You will notice that there is no radiation off the end of the aerial, so although the aerial is radiating signals outwards from itself, the effect now is to concentrate the radiation to the left and right of the page. No radiation is taking place towards the top or bottom of the page. Such an aerial is therefore directional. In order to cover all points of the compass, the horizontal dipole has to be rotated.

It is possible to carry this aspect further and concentrate all the radiation in one direction. In this way a beam aerial is created. This will certainly have to be rotated to cover all possible contacts. If this seems to be a disadvantage, the point to remember is that all the power available is being concentrated in this one direction. Since it is usual to be in contact with only one station at a time, power going in other directions is wasted. This concentration of power into a beam gives considerable gain, compared with a simple dipole. This gain costs nothing in terms of power at the transmitter, so it is worthwhile. It is rather like a torch bulb. When simply connected to a battery its light is not very impressive. Concentrated into a beam by a torch reflector the light is intense in the area where it is aimed, but it must be moved around to obtain full coverage. So it is with the beam antenna.

Any aerial, resonant or not, in the path of a radio wave, will have small RF currents set up along its length, these same currents will cause some of that energy to be re-radiated, acting as a transmitting aerial. If the wire is a resonant length, and at the same time unconnected to any load, most of the energy will be re-radiated. In the beam antenna or Yagi, this property is made to work to advantage. Fig. 65a shows a simple three-element beam with the dipole itself, an element called the reflector, and an element called the director. These extra elements are known as parasitic elements, and there may be several more directors in more complex antennas. Obviously, the frequency at which the aerial is resonant will determine the length of the elements and therefore the number of them that can be

ANTENNAS

practically accommodated. For twenty metres an aerial is
unlikely to have more than three elements. A 70cm aerial
could easily have 18 or more and still be stable enough to
be supported.

The spacing of the parasitics in Fig. 65b is critical since it
determines whether the element is a director or reflector. If
the distance of the reflector from the dipole is correct, then
the pulse it receives will be re-radiated out of phase with the
pulse following, and thus they cancel. If the spacing of the
director is also correct for its purpose then the re-radia-
tion here will be in phase and thus aiding the signal. In this
way the signal is enhanced in the forward direction and
attenuated in the backwards direction.

Similarly, the beam will respond to signals from the forward direction and enhance them but will not respond to
signals from the rear. Many designs and variations on this
simple theme can be found, but the two most important
characteristics are: front to back ratio, in other words how
much is going out in the forward direction compared with
that radiating in the backward direction, expressed as a
ratio (e.g. 6:1). The other charactertistic is forward gain.
An aerial itself does not amplify a signal, but a beam will
produce greater radiation in the forward direction than a
simple dipole, it is this comparison that the forward gain
figure refers to.

Gain in antennas and amplifiers is expressed in
dB—decibels. This is a means of expressing a change, up or
down, of power levels. The basic unit, the Bel, is a change of
ten times. This is a rather large unit, and we use the tenth
of a Bel, the decibel. A 1dB of gain is an increase of just
over a quarter. In other words, 1 watt increased by 1dB
would become 1·26; 3dB gain is double—1 watt becomes 2
watts. Therefore 6dB is a power gain of 4 and 10dB, as we
said, is an increase of ten times. A complete list of dB gain
and attenuation is available in most reference books.

Before leaving the beam aerial, there is one feature which
should be mentioned. The dipole element is often folded or
constructed of parallel rods joined top and bottom. The
reason is that the normal impedance of 70 Ω at the centre
of a dipole is reduced to about 20 Ω when parasitic elements

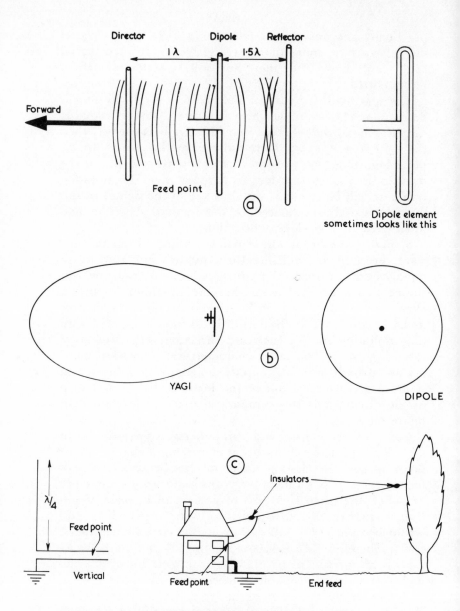

Fig. 65. Further aerial variations
      a. 3-element yagi
      b. radiation pattern of yagi
      c. Quarter wave vertical and horizontal

are added. Folding the dipole increases this by a factor of 4, and so we finish up with about the right value again. Sometimes a form of transformer is used to obtain the correct matching.

On VHF bands, and the HF bands down to twenty metres, it is practical to use the dipole and the beam. For the lower frequencies the length of the elements precludes this type of antenna. The quarter wave vertical is very popular and with modification can be used even at 160 metres, where a quarter wavelength is 40m. or about 138ft. The $\frac{1}{4}$ wave vertical, or Hertz aerial, behaves electrically like half a dipole, with the missing half being reflected in the earth or ground plane. The impedance at the base of this type is about 50 Ω and an efficient earthing system is necessary for satisfactory operation. An ineffective earth means that resistance is present between earth and transmitter and results in power being dissipated in this resistance, which in the case of the quarter wave, is not a radiating element, and is thus lost. For low frequencies, the long length of the wire need not go vertically skywards, but may be folded over and lie largely horizontal—see Fig. 65c. It is then more often called simply an end-fed antenna.

For many amateurs the possibilities of more than one aerial are limited and yet they would like to operate on more than one band.

Separate VHF aerials are not usually difficult to accommodate because of their short length. For the HF bands the multiband aerial is a means of making one antenna serve more than one band. The long wire can be made so that its length—from the transmitting point to the far end—is a quarter wavelength at the lowest band required, say 2MHz. Such a wire would need to be 130ft. long. End fed, it will present a low impedance to the transmitter. Coupled through an ATU (antenna tuning unit), this low impedance can be adjusted so that it perfectly matches. The same aerial can be used on a higher frequency when the feed point becomes a high impedance. This can be matched to the antenna by an ATU which will transform the high impedance down to a low one (see Fig. 66). Very often the two ATU configurations can be combined in one. Switches

141

rearrange the components from series to parallel and also select different tappings on the coil to vary the resonant frequency.

Another way to make a multiband aerial is with a wave trap, thus creating the trap dipole or trap vertical. Basically, this consists of an antenna whose length is that of the lowest frequency band to be operated. A trap is then inserted at the length of the next highest frequency, which will effectively shorten the antenna to the position of the trap (*see* Fig. 66c). Here the length A corresponds to the lower frequency wavelength and length B the higher frequency. The trap is tuned to be resonant at this higher frequency. Being a parallel tuned circuit, and thus having high impedance, it acts like a switch, cutting the antenna at this point. When fed with a frequency at the lower wavelength, however, the coil has a low impedance and couples in the extra length of aerial.

This trap principle can be extended to make a five band system. Such an aerial is a compromise, of course, and some efficiency is sacrificed, but for many operators it is the only means of achieving coverage of all amateur frequencies.

The construction of antennas is a most important factor in their efficiency. Since they are outside, and subject to all weathers, corrosion is a potential enemy. With trap aerials especially, the tuned circuits must be fully weatherproof, since any deterioration in this area can cause serious problems of interference. Any breakdown or change of values in the trap means that it is no longer functioning at the frequency it should, and the aerial as a whole may be resonant at a frequency differing from that being fed to it.

This type of fault can cause interference by generating signals outside the amateur bands.

The components used in the construction of antennas and traps should be sturdy and capable of withstanding high voltage. Aerials should be suspended as far as possible from trees and buildings, especially overhead electricity wires.

Obviously the further they can be sited from TV and radio aerials, the less chance there is of causing interference.

As with transmitters and receivers, a study of the amateur

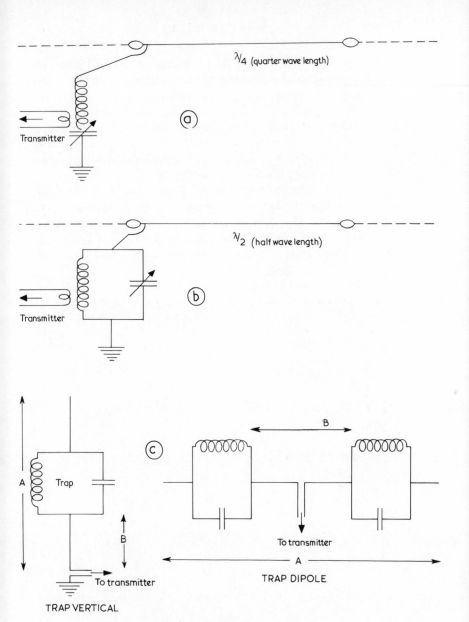

Fig. 66. Multiband aerials
    a. Aerial tuning unit — series tuned
    b. Aerial tuning unit — parallel tuned
    c. Wave traps

press will reveal a bewildering range of antennas each with its own name and claim. It is exceptionally difficult to evaluate an aerial by reading about it, since locations vary enormously in their siting, the nature of the soil, the geographical position and so on. You can be sure of one thing, antennas will often defy antenna theory. Every amateur will have his own version of a story where the proverbial piece of wet string has outperformed some complex and highly sophisticated antenna.

Be cautious when evaluating an aerial. Try to obtain unbiased reports from other amateurs and listeners. If possible, study the layout of a successful station whose location is similar to your own. Remember the 'gain' of an aerial is meaningless unless compared with a reference of some sort, i.e. a dipole.

Finally, it is not the power the transmitter puts into the aerial that counts, it is the power leaving it!

Chapter Fourteen

TRANSMISSION LINES

In some situations the antenna can be coupled directly to the transmitter. The walkie-talkie or hand-held transceiver, for example, has a quarter wave antenna, the base of which is the feed point, and is located directly at the output amplifying stage. We have seen the end-fed aerial whose feed point can be brought into direct contact with the transmitter output at the aerial tuning unit.

In many stations, however, the aerial and the transmitter will be sited some distance apart, the one indoors, and the other high up outside. The dipole for instance should be mounted high and as far away as posible from buildings and trees. The purpose of the feeder, or transmission line, is to connect the two and convey power from one to the other.

It is not the purpose of the feeder to play any active part in the function of the aerial. It should therefore not modify the signal nor radiate it, nor indeed receive any signal itself. It should simply transfer the power without loss to the radiating element. It has a similar function to the transmission shaft in a car or lorry. Here the power generated by the engine is transferred by the shaft to the road wheels. Many people refer to transmission lines, such as the one connecting the TV set to the roof aerial, as 'aerial wire'. This is simply not so. The feeder is playing no part (if the installation is a proper one) in the received signal, but simply transferring the power generated in the aerial by the intercepted signal and conveying it to the set. One could use ordinary wire as a feeder to connect the aerial to the transmitter or receiver, but at the same time as it was transferring power it would also be radiating and dissipating power from itself, so it is clearly not suitable as a transmission line.

The feeders in common use today are of two types. One is twin feeder — parallel wires, usually encased in a plastic ribbon, in which each wire has an equal potential with respect

to earth. This is known as balanced feeder. The other type
is coaxial feeder, again with two conductors, but in this case
one of them is coaxially wrapped round the other. Again,
plastic spacing and covering maintains the dimensions and
protects the wire.

In coaxial transmission wire the outer conductor is usually
at earth potential, so it is known as an unbalanced feeder
(*see* Fig. 67a).

The most common of the two in use today is the coaxial
type. Either, however, when fed with an RF current, will
exhibit a characteristic impedance. This property arises out
of the diameter of the wire used in the construction and the
spacing either between the two wires or between the wire
and the outer braiding. The conductors can be considered
as a series of inductances running the length of the feeder,
together with the capacity existing between the conductors
(*see* Fig. 67b). The value of this characteristic is most commonly 75 Ω or 50 Ω in coaxial cables since this matches the
radiation resistance of dipoles and quarter wave aerials.
Twin feeder is available in 75 Ω or 300 Ω values. The impedance is distributed over the whole length of the line.

A signal current generated at one end of the line will be
transmitted throughout its length, and the power dissipated
in a load connected at its further end, provided the output
impedance of the generator, the impedance of the feeder
and the resistive load are all of the same value.

As the waves travel along the line, there is a phase lag
due to the inductance. The varying RF signal charges up
each C through L, Fig. 67b. The inductance preventing the
capacitor from charging to its maximum until a fraction
later in each succeeding LC pair. The speed of the wave is
thus delayed by a small amount compared with its speed in
free space. The ratio of these two speeds is known as the
velocity factor of the line. For coaxial cable this is approximately 0·67 to 0·86, and for 300 Ω twin about 0·85.

When the line is correctly terminated, all the power is
dissipated in the load. This is the objective when it is an
aerial that forms the load. The effect of an incorrect ter-
mination or mismatch, is that some of the energy will be
reflected back along the line. The two extremes of mis-

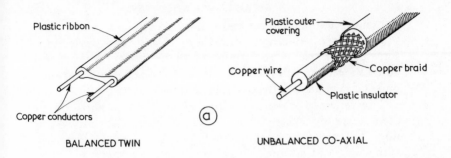

BALANCED TWIN   UNBALANCED CO-AXIAL

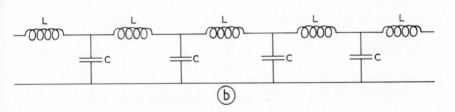

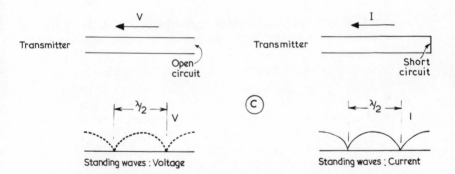

Fig. 67. Transmission lines (feeders)
    a. Balanced twin and unbalanced coaxial
    b. Electrical equivalent of a transmission line
    c. Mismatched lines. Open circuit with voltage reflection and
       short circuit with current reflection

match are an open circuit and a short circuit.

From an open circuit line there is a voltage reflection, since this represents an infinitely high resistance. A short circuit line on the other hand has an infinitely low resistance and the current is reflected (Fig. 67c).

The reflected waves meet the forward waves and add to them to produce a series of standing waves back along the line. The presence of the standing wave nodes is detectable with the aid of a loop of wire connected to a meter or bulb. Passing the loop along the feeder will cause the bulb to light, or the meter to indicate that a current has been induced in it.

Standing waves are undesirable for several reasons. Firstly their very presence indicates that not all the power is being dissipated in the load, which, if it is the aerial, is where it should be dissipated. Again, standing waves mean the presence of current on the outside of the feeder and almost certainly radiation will take place from here as well as from the aerial. This is likely to cause interference since the feeder may pass close to other receiving aerials or systems, and will offset the advantages of any directional properties that the terminal antenna may have.

In addition, reflected currents arriving back at the transmitter impose stresses on it for which it was not designed and damage may ensue.

Standing waves then are a means of indicating how efficiently the aerial is matched to the line. A more practicable means of measuring this factor than the one mentioned above is the reflectometer or standing wave ratio meter. It is a device which samples the current both forward and reverse over a short length of line, usually coiled up within the meter cabinet. This is then displayed either on two separate meters simultaneously or on a single meter switched from one state to the other. The device is simple to use. Power is fed from the transmitter along the line to the terminal load.

An adjustment in the meter allows the forward current to be set so that it just causes full scale deflection. The reverse reading is then taken. The scale is calibrated in figures representing the SWR, i.e. the optimum is 1, with no reflected

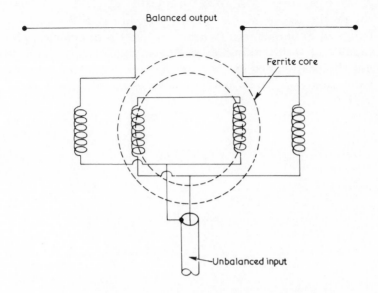

Fig. 68. BALUN – Balanced to unbalanced transformer

current. The worst condition is infinity, a complete mismatch.

A completely perfect match is almost impossible to obtain, but a ratio of 1·5:1 should be aimed for. Many transmitter manufacturers specify a maximum figure which should not be exceeded in order to protect the output stage. Some designs are protected by automatic shut-down circuits, should the SWR exceed a given level.

Ideally, the SWR should be measured at the aerial feed point, but as this is seldom possible, allowance must be made when reading the meter. Losses on the feeder itself will tend to make the SWR reading more favourable than actually is the case. The further it is away from the antenna, the more inaccurate the reading will be.

One final point about matching is this. Many aerials are of the dipole configuration, which is balanced, and fed by coaxial cable which is unbalanced. If connected up directly in this way a state of imbalance will exist. The outer 'earthy' sheath of the cable will become live with RF currents and again possibilities of interference arise. The answer is to

149

couple the feeder to the aerial by means of a BALUN—a
balanced to unbalanced transformer. Such a device has to be
capable of withstanding all weathers, and for this reason is
usually encapsulated in resin, having been wound on a ring of
ferrite material. An HF balun is shown in Fig. 68.

Chapter Fifteen

PROPAGATION

We come now to the very nature of radio waves, and how propagation takes place. It is very difficult to describe something which cannot be seen or felt. Rather like the current we described earlier, it must be taken on trust. At least with current there were the circuits through which it flowed, which could be shown and defined. Electro-magnetic radio waves do not require a medium to flow through. This form of energy is completely self sustaining, but like the bark, which still occurred when the dog was removed, the presence of radio waves can be proven by the manifestation of their end product, i.e. radio programmes, television, etc.

The earth is continuously in the path of electro-magnetic radiation arriving from outer space, originating from the sun and the stars. All these radio waves have different frequencies, and have been grouped into bands according to their characteristics. Light, for example, is the most familiar one. The frequencies within this band are extremely high 200×10^{12} MHz. (200 million million oscillations per second approx, for blue light). Individually they give us the different colours, collectively they appear as white light. There are other bands, ultraviolet and infra-red for example are light waves invisible to the human eye. Then there are X-rays and gamma rays which are higher still in frequency and which can be dangerous.

The exposure to dangerous rays from the sun is not usually harmful since the atmosphere acts as a filter, but sunburn is an example of tissue damage that can arise from radiation. When we are X-rayed at the hospital or dentist the amount of radiation is strictly controlled and kept well within safe limits. There is no danger from the frequencies and powers that amateurs use to people in the vicinity of their aerials.

It is to the radio wave bands that we now turn and the two main bands which interest the majority of amateurs are the

151

HF and VHF bands. The electro-magnetic nature of all these waves is identical.

Radiation is generated in a transmitter by causing a changing current at radio frequency to oscillate backwards and forwards along an aerial wire. This will create a magnetic field around the wire that is at right angles. These fields radiate outwards, mutually reinforcing one another. If the aerial is vertical, as in the diagram, then the magnetic field will be horizontal and the electric field vertical, a horizontal wire would produce the opposite effect. The polarity of a radio wave is that of the electric field, so our example in Fig. 69a would be described as vertically polarised.

The wave moves outward at great speed 300×10^6 metres per second, about 186,000 miles—the speed of light. Light from the sun takes about eight minutes to reach us. The fact that all stars (other than the sun) are light years away gives some idea of the incredible distances encountered in space.

The radiation from a vertical aerial will be omnidirectional, that is spreading equally in all directions. The two components of the wave approaching the observer are diagrammatically depicted as a grid (Fig. 69b.) If a wire is placed in the path of such a wave the alternating electric and magnetic fields will induce a current within it, the frequency of which will be the same as that in the transmitter. This is the basis of propagation.

In practice, of course, there are at all times thousands of radio waves, both man-made and naturally occurring, inducing into any conducting wire, whether it be clothes line, coat hanger, or TV aerial, myriads of tiny alternating currents superimposed one on another. The basis of communication is to sort these frequencies out and make use of them.

The part played by the receiving or transmitting wire in all this has been discussed in the chapter on 'Antennas'. This is the term used to describe aerials in the communications field.

The speed at which radio waves travel is constant (in space), and as it is an oscillating quantity, a wave will travel a fixed distance during the period of one alternation. This distance is called its wavelength, and it follows that the faster the alternation is completed, i.e. higher frequency, the shorter the distance covered. Hence the higher radio frequencies are

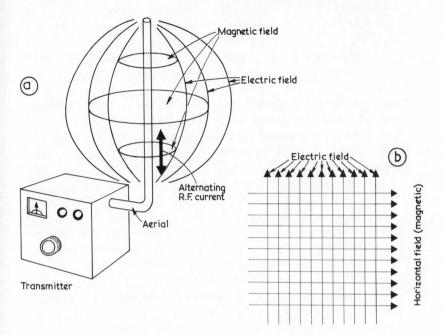

Fig. 69. a. Radio waves being generated: alternate magnetic and electric
fields are emitted outwards from the aerial
b. Radio waves approaching the observer. The signal is vertically
polarised. The next half cycle the arrows would reverse

called short waves and the lower frequencies long waves.

This relationship between wavelength and frequency can be
expressed in the formula:

$$\lambda = \frac{300 \times 10^6}{f}$$ where λ (lambda) is wavelength
in metres.

f = frequency in Hz.

or: $f = \dfrac{300 \times 10^6}{\lambda}$

The three types of propagation by which a radio wave may
travel are: ground wave, sky wave (ionosphere), and
tropospheric wave. Radio waves can also travel through solids,
and as we have seen, along conducting wires. Solids will cause
absorption and attenuation of the wave to a varying degree,

depending on the nature of the solid and the frequency of the wave.

Ground wave propagation is where the transmitted radio wave follows the contour of the earth's surface. It is the mode of propagation by which we receive our long wave and medium wave broadcasts. It is effective up to about 2MHz, and this includes the amateur 'Top Band'. Beyond this attenuation of the signal by trees, terrain, buildings, etc. increases. At 28MHz (the ten metre band) ground wave radiation is perhaps a few miles.

Sky wave propagation is by reflection from the ionosphere. It is the main means of propagation for amateurs in the bands up to 30MHz, and is a rather complex subject. Frequencies behave differently at night from their day-time behaviour, and differences occur seasonally, and also over an eleven-year cycle attributed to sunspot activity. Other sporadic factors, like weather, solar storms, etc., will also affect propagation.

The ionosphere forms part of the atmosphere, it is a layer of ionised gases lying above the stratosphere and the troposphere. The atmospheric gases in the ionosphere, which starts about thirty miles up, are very rare. The molecules, mostly nitrogen and oxygen, are ionised by the sun's rays. That is to say radiation from the sun strips electrons from their atoms, leaving charged particles called ions.

Radio waves, up to about 30MHz, can be reflected by these ionised layers and returned earthwards. Because of the rarified nature of the upper atmosphere, wavelengths shorter than this penetrate the layers and continue on into outer space.

The ionosphere is subdivided into three layers, D., E. and F., one of which, the F. layer, is divided again into two, F1 and F2.

Fig. 70a shows a simplified sketch of the layers. These are not hard and fast distances of the height of the layers, but serve as a guide. There is no sudden transition from one layer to the next, and the degree of ionisation varies at different times. In general, the D layer is weakly ionised, and plays little part in the reflection of signals. It can, however, in some circumstances absorb signals, thus attenuating them and inhibiting their reflection by the E or F layers. The E layer is more strongly ionised, but this diminishes during the hours of darkness. The F layers are most strongly ionised and contribute to reflection throughout the night.

It is because of these reflecting properties of the ionosphere that we are able to bounce signals back to earth at a point far beyond that which is reached by the ground wave. If conditions are good, this reflecting process can take place again and again from earth to ionosphere and back, and world-wide communication can be achieved (*see* Fig. 70b). Between the end of the ground wave and the point at which the skywave returns is the skip distance. Between these two points is a dead zone, where no appreciable signal will be heard. A change in the skip distance, which itself can be attributable to a number of factors, is a major cause of fading and is often quite unpredictable.

Two definitions are used in determining the performances of the ionised layers as they change due to season and sunspot activity. The MUF, maximum usable frequency, is the highest frequency that will be reflected at a given angle to the ionised layer. Frequencies above this penetrate the layer and escape into space.

The Critical Frequency is the highest frequency that will be reflected back from a vertical radiation (*see* Fig. 70b.). These frequencies will vary according to the degree of ionisation. This in turn depends on the time of day and night, and the season, because these factors determine the amount of solar radiation received. Sunspot activity also affects the degree of ionisation. There is an eleven-year sunspot cycle with regular maximum and minimum activity occurring at regular intervals. The ten metre 28MHz band is the most dramatically affected. At the minimum period this band is virtually dead, at the maximum period, five watts will put a signal round the world easily. The difference is that great. At the time of writing, early 1980, the cycle is just about at its maximum, it will sink into a minimum in 1985/6 and will be at a maximum again in 1991.

Because these functions are not constant, and the layer's reflecting surface not a fixed and definite point from which the signal emanates, its strength at the receiver point may vary enormously.

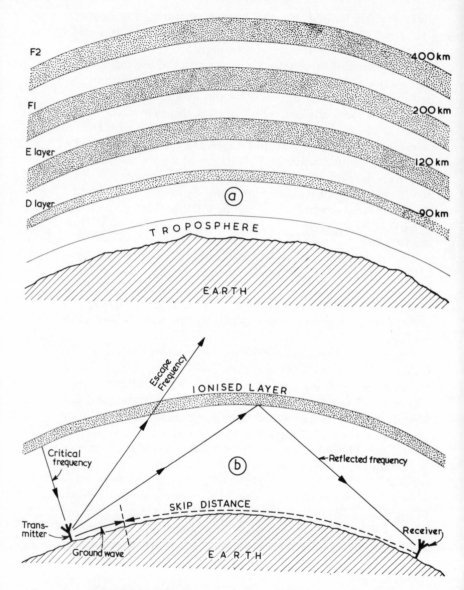

Fig. 70. Layers of the ionosphere and reflection of radio waves

This, together with the skip change, is another cause of fading. Sometimes the receiver is on the edge of a ground wave and also in range of a skywave. In this case, because of the different distances they have travelled the signals may be first in phase, and aiding, then out of phase and cancelling, which is yet a further cause of fading. Thus the need for really effective AGC action in a communications receiver is apparent.

Fade out is a somewhat different phenomenon, and may last for several hours, or even days. It occurs when ionospheric propagation is disturbed by abnormal radiation from the sun. Intensive ionisation of the D layer can cause it to absorb signals before they have the opportunity of reaching the F layers.

So far we have considered reflection in the HF bands up to 30MHz as a way of propagating radio waves. At frequencies above this, for example the two metre and seventy centimetre bands, the frequencies will pierce the ionised layers and escape into space. So for the main part communication in these bands 144MHz and 430MHz is by ground wave. To be effective, the transmitting and receiving stations should be in line of sight or nearly so, as the signals are rapidly attenuated by buildings and other masses. In any case, the horizon will be the limit, since at these frequencies the wave tends to travel in a straight line.

However, propagation over longer distances is possible by refraction. Refraction is a bending of the waves through a shallow angle. This happens when looking at an object in water. The light waves bend, and the object seems displaced. The way refraction can help is shown in Fig. 71. This refraction depends on a change of the dielectric constant (K) in the atmosphere at the height of a kilometre or so. Changes in K depend on temperature, humidity, etc. In certain weather conditions layers of different temperatures of air become trapped in a sort of sandwich and it is through these layers in the troposphere that refraction takes place.

When these conditions arise, and they are not really predictable, except in the short term, we say there is a 'lift', and the VHF bands become very active. While these 'lifts' give amateurs the chance of working stations much further afield than usual, the same conditions are causing TV signals to be

157

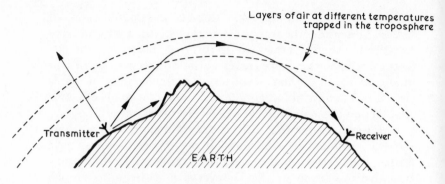

Fig. 71. Tropospherical propagation by refraction

propagated further, so during these periods there is very often interference on the TV channels from Continental and other distant transmitters.

In the next chapter we shall discuss the operation of Repeaters and Satellites in the amateur bands. For the moment, suffice it to say that they are another means by which VHF propagation, can be extended beyond the line of sight range.

There are other means of propagation, such as moon bouncing (reflecting signals from the moon), but these are rather specialised and cannot be considered as an everyday means of communication for the average amateur.

In conclusion, here is a summary of the approximate distances that can be obtained by radio signals in the various amateur bands:

1·8 MHz band.	50 miles ground wave.
	Several hundred miles possible at night.
3·5 MHz band.	3-500 miles.
	Between six and seven thousand at night.
7·0 MHz band.	Similar, but limited ground wave 20 miles.
14·0 MHz band.	World wide, but very limited ground wave of about ten miles.

21·00 MHz band.	Similar, but more affected by conditions, e.g. day, night, season, sunspot cycle. Shuts down at night.
28·00 MHz band.	Varies from inactive to extremely good, depending on the eleven year sunspot cycle.
70/144/430 MHz bands.	Mainly line of sight or via repeater or satellite. Greater distances possible, several hundred miles when unusual tropospheric conditions occur.

Amateurs also make use of moon bounce; Troposcatter (ionised clouds within the troposphere); Auroral reflection from magnetic storms in the polar regions; Meteor trail scatter, a trail of ionisation of short duration in the presence of a meteor.

Chapter Sixteen

REPEATERS AND SATELLITES

The limited range available from the VHF and UHF bands can be artificially improved by means of the repeater and the satellite.

Since the early 60s a series of satellites designed by amateurs for amateurs has been launched. These are called OSCAR 1, 2, 3, etc. —from 'Orbiting Satellite Carrying Amateur Radio'. Within the satellite is receiving and transmitting equipment powered by solar cells. A solar cell is a device covered with a substance which produces a current when exposed to solar energy—sunlight.

The satellite acts as a transponder, receiving signals in one amateur band and re-transmitting them in a different one. At present they are: —

432 MHz - 144 MHz
144 MHz - 432 MHz
144 MHz - 28 MHz

Oscar satellites will transpond any from of modulation and will work from a transmitted signal of about 100 watts ERP (effective radiated power). Bearing in mind the gain available from a directional antenna, 100 watts ERP could be obtained from a 10 watt output into an array with 10dB gain.

Since the satellite is orbiting, its azimuth elevation and time of orbit must be known in order to be able to beam the aerial in the correct direction. This data is available in the form of predicted orbits, to those who are interested in this aspect of the operation.

Obviously, the line of sight range of VHF and UHF transmission is vastly extended when carried out through a satellite transponder.

A more limited means of extending the range of such frequencies intended for local FM working, is the repeater.

This device, again a transmitter receiver, is located on high ground which gives it good line of sight coverage over the surrounding area. It is usually unmanned and automatic and responds to one fixed frequency (channel) only. The signal is retransmitted as before, but this time in the same band and with a frequency shift.

The two metre repeater will accept a signal in one of the repeater channels R0 to R7, convert them up by 600 kHz and retransmit them, The two metre transceiver must have a 600 kHz shift between its transmit and receive frequency in order to operate in both the input and output channels of the repeater.

For example:

To transmit on R0 input is 145·000MHz.
To receive on R0 output is 145·600 MHz.

The 70 cm band has a 1·6MHz repeater shift and this time it is downwards in frequency between input and output.

The repeater transmitter is not operational continuously and is only activated when the repeater receiver responds to a transmission at the correct frequency and containing a short burst of modulation lasting half a second and consisting of a tone of 1750Hz. This timed tone is recognised and used to activate the transmitter. It is included to help prevent spurious operation of the repeater. Once the tone has been recognised, a timing circuit commences which limits the period of transmission to 80 seconds or so. This is to keep transmissions short so that maximum use may be made of the repeater, since only one channel is available for use in any one location.

Repeater channels and location and other information is published from time to time by the RSGB.

Chapter Seventeen

MEASUREMENT

Under the terms and conditions of his licence, an amateur must be able to make certain measurements. Voltage and current measurements can give information about the power being developed by a stage and thus ensure that the transmitter is operating within the power limitations set by the Home Office. The licence also requires a means of accurate frequency measurement to be available for establishing that radiation is within the permitted amateur bands. Many stations, too, use an instrument called the oscilloscope. This is a device for displaying signals in two-dimensional form on a changing display like a small TV screen, called the Cathode Ray Tube (CRT).

First let us look at voltage and current. The basis of most instruments for measuring these is the moving coil meter. This consists in simple terms of a coil of wire set on bearings and lying in the field of a permanent magnet, *see* Fig. 72a. This is the electro-magnetic effect again. When a current is passed through the coil a deflecting force is created by the magnetic field, and the coil starts to rotate. Its rotation is resisted by the coiled springs. A pointer attached to the coil can be used to indicate on a scale the amount by which it has rotated. This in turn is proportional to the current flowing.

The scale therefore can be calibrated in amps or milliamps or whatever is appropriate.

In the instruments we use to measure the voltages and current encountered in amateur radio equipment, a very sensitive movement is required. A meter whose Full Scale Deflection (FSD) can be caused by a flow of only 50μA will be able to measure extremely small currents, since one or two microamps will give a detectable deflection on the scale.

This is very useful in today's circuits with transistors and diodes which work at very low voltages and currents. Most multimeters have only one movement, and yet offer a range of measurements, often up to a thousand volts and several amps, by using shunts and multipliers.

First the shunt: this is a resistor in parallel with the meter movement (*see* Fig. 72b). This offers a path for some of the current to flow through, and by making R much smaller than the resistance of the movement itself, the larger part of the current will pass through the shunt. In this way, several current ranges can be dealt with, a different value of Rs being switched in for the purpose. For example, consider a meter with a basic FSD of 1 amp and whose movement has a resistance of 10 ohms. the voltage across this meter at FSD would be (from Ohm's law) E=IR=10 volts.

Suppose now that the meter is required to measure a current ten times greater, 10 amps. Then the shunt resistor Rs would have to be of such value as to bypass the excess current, that is 10 - 1 = 9 amps. Knowing the voltage drop at FSD we can again work this out from Ohm's law.

$$Rs = \frac{E}{1} = \frac{10}{9} = 1 \cdot 111 \text{ ohms.}$$

Shunt values are always odd quantities like this, and therefore special resistors have to be made for these applications. In electronic measuring instruments, we are more likely to encounter the $50\mu A$ movement mentioned earlier, and if this was required to read, say 50mA, the shunt value required would be $0 \cdot 01001\Omega$. This again shows the special nature of this component, very low ohmic resistance and an odd value.

To measure voltages, a slightly different technique is used. The basic movement can only respond to current flow so this has to be interpreted in voltage levels. Supposing the basic $50\mu A$ movement is required to measure a PD of 10V at FSD. Referring to Fig. 72c a resistor is added in series with the meter. This time it is known as a multiplier. The value is arranged so that together with the meter resistance a PD of 10 volts between points A and B will cause a current of $50\mu A$ to flow.

The total resistance, therefore, between A and B is: —

$$R \text{ total} = \frac{E}{1} = \frac{10}{50 \times 10^{-6}} = 200,000\Omega$$

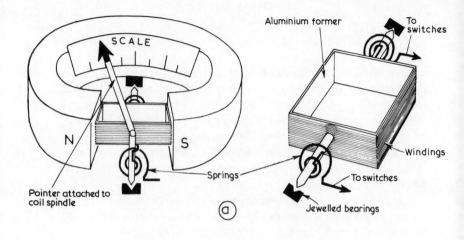

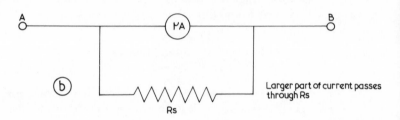

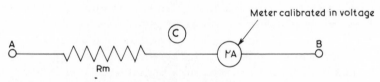

Fig. 72. Basic moving coil milliammeter
   a. Coil detail
   b. Meter shunt
   c. Meter multiplier

The meter resistance has to be deducted from this, so the value of the multiplier is: —

$$Rm = 200K\Omega - 10\Omega = 199990\Omega$$

Again a special value.

Meters with the capability of measuring several ranges of

current and voltage are known as multimeters. There are numerous types available, most of which include one further facility—that of resistance measurement. This is how it works. The meter itself, as already stated, can only measure the presence of current. So in this case current is supplied from an internal source, usually a small battery of the type used in torches.

The meter is first calibrated, that is to say in the resistance measuring position the internal battery is brought into circuit and its current output is adjusted by a variable resistance. By connecting the two measuring probes of the meter together, all the output of the internal battery flows through the movement and is adjusted for FSD. On the resistance scale this point is marked zero. If now the probes are separated and connected to an unknown resistance, the current from the battery will be less and the meter will indicate part way along the scale. Knowing the voltage which will cause the FSD of the movement alone, decreasing current can be interpreted on the meter scale as values of resistance (Fig. 73a). Thus the ohm range on a meter always appears back to front. Zero on the right, infinity on the left. The scale is logarithmic (not evenly graduated), and becomes less accurate as the resistance goes higher. Many meters have several ohm ranges and more than one battery.

The scale on a multirange instrument must have at least two ranges.

As we have seen, resistance calibration will be opposite to current and voltage. So as not to create too much confusion the various voltage, current, and ohm ranges each has its own basic calibration, and then to read a higher range, the basic one is mentally multiplied by 10, 100, or whatever factor is applicable. For lower readings the scale may have to be mentally divided by some factor.

Finally, there is AC to be considered. Alternating voltages or currents are first rectified, and the resulting pulsed DC fed to the meter movement. The scale is calibrated to read the RMS value of the applied voltage or current.

When using a multimeter the golden rule is to start off on the highest range and switch progressively down until a suitable meter deflection is obtained. Many instruments have

165

an automatic cut out in the event of an overload, but it is possible to defeat this, and if a high current or voltage is applied when the instrument is set on a low range, expensive damage can be caused in a split second.

Frequency Measurement. The first and simplest frequency measurement device we shall look at is the Absorption Wavemeter. As the name implies it derives its power by absorption from the circuit under test. Fig. 73b shows a basic circuit. The coil L and capacitor C form a tuned circuit. The capacitor is connected to a calibrated scale and the coil L is usually one of a series which can be interchanged to give a wide coverage of frequency. The coil will be extended from the instrument so that it can be brought into close proximity with the circuit under test. Sometimes a short rod antenna is used. The tuned circuit under test is activated and the wavemeter is coupled to it as closely as possible. Currents from the tuned circuit are induced into the wavemeter circuit and the capacitor is rotated. Resonance is indicated by deflection of the meter. Such an instrument is very useful in testing for the presence of harmonic radiation from a transmitter. Apart from its simplicity it has the advantage of direct frequency calibration. The reading is unambiguous.

Such an instrument does not give an accurate frequency readout. One range may cover typically $1 \cdot 5 - 4 \cdot 0$ MHz, so within this range the frequency is only approximately indicated. However, this is perfectly adequate for the rough checking of a frequency, and for detecting the presence of harmonics. It can also be used to support the readings of a more accurate device such as the Frequency Counter.

The frequency counter is a very accurate, and far more complex piece of equipment. The absorption wavemeter is so simple that the majority are probably home made, or 'home brewed' as amateurs like to say.

The digital frequency counter is far more likely to be commercially produced. It works by counting the number of pulses occurring in a fixed time. This fixed time is a very accurate fraction of a second, itself produced by a crystal based oscillator. The information thus derived is displayed in digital form, rather like that in a calculator.

A 'sniff' of the signal to be measured is fed to the DFM

166

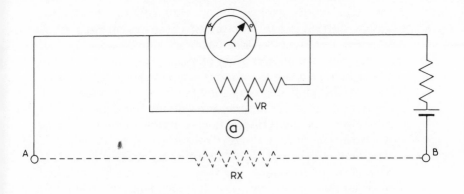

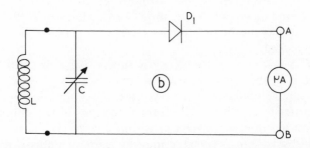

Fig. 73. a. Resistance measurement. Rx is the unknown value. A and B
are shorted together and Vr adjusted to give FSD (zero). Rx is
then connected between A and B, giving reduced current
flows. The scale is graduated to show this deflection in ohms
b. Absorption wavemeter

usually by means of a short length of pick-up antenna. The
sine wave nature of this signal is converted into pulses internally, counted and displayed to an accuracy of just a few
cycles.

Digital displays are becoming increasingly popular, not only
on frequency meters like this, but also incorporated into transmitters, receivers and even volt, ohm, and current measuring
instruments.

The advantages of these devices is that they take only an
infinitesimal quantity of power from the circuit they are

measuring. In the case of the moving coil multimeter for example, current has to be diverted at the expense of the circuit under test, to motivate the meter movements. This of course causes a small inaccuracy of measurement.

Many stations make use of the crystal calibrator for accurate frequency checking. By comparing the frequency of one of the harmonics of an accurate crystal oscillator the calibration of the receiver can be verified. It is a simple matter then, to calibrate by comparison the transmitter with the receiver.

The heterodyne wavemeter works on a similar principle. It has an accurate oscillator, the frequency of which is checked against an internal crystal. The desired frequency is set, and its output is compared with the output from the transmitter.

The transmitter would normally be coupled to a dummy load for this sort of check, to prevent radiation. A short pickup antenna on the heterodyne wavemeter receives the signal to be checked.

The transmitter frequency is slowly adjusted until a beat note is heard. At zero beat the transmitter will be at exactly the same frequency as that set on the heterodyne wavemeter. In order to use this method, the frequency of the signal being measured must be known to within a few kHz. This is where the absorption wavemeter may be used to approximate the frequency, the heterodyne wavemeter to pinpoint it. The accuracy should be within 1 kHz.

Reference has been made to the dummy load, which should be used when any tests or checks are being undertaken. It enables all the functions of the transmitter to take place, but instead of the power output being radiated it is dissipated in the form of heat. The load takes the form of a resistor whose value will be the same as the transmitter output impedance, usually 75Ω or 50Ω. It must also be of suitable rating to dissipate the specified power output. This could mean as much as 100 watts. Most high power resistors are wound with resistance wire. This type however is unsuitable for dummy load purposes since the coils of wire have inductance and this in turn creates reactance which will vary with frequency. So in order to present a constant resistance at all frequencies the load resistor must be of the carbon type.

The licence limits the amount of power that an amateur

may use. This power is defined in two ways. First the maximum DC power input. On most bands this is 150 watts. On the 160 metre band it is only 10 watts and in the 4 metre band it is 50 watts.

DC input power is defined as the total direct current power into the device energising the aerial. In the case of AM and FM transmitters where a constant current flows in the output amplifier, this is simply voltage multiplied by current. From the formula, power $= E \times I$ (watts).

With an SSB transmitter, however, there is no constant carrier and the output current is intermittent, occurring only when modulation takes place. Without a steady current the above method cannot be used to measure power. This brings us to the second method of defining power.

The Peak Envelope Power (PEP). Under linear operation the peak power is limited to $2\frac{2}{3}$ times the appropriate DC power level. This means that 150 watt DC becomes 400 watts PEP, 50 watts becomes 133 watts PEP, and ten watts becomes $26\frac{2}{3}$ watts PEP.

The recommended way of determining this power is rather complicated and requires the use of an oscilloscope. The method is outlined here for those who wish to follow it through. Many commercial transmitters, however, have maximum outputs considerably less than the permitted 400 watts PEP and it would not be necessary in these cases to go through this procedure since operation would always be well within the permitted limit.

The oscilloscope is a device for displaying the two dimensions of a signal, its frequency and its amplitude, on a screen very much like a TV screen.

Power Measurement of RF amplifiers in J3E (single sideband). The RF PEP output, under linear operation, of a J3E signal should not exceed that of an A3E transmitter operating at maximum permitted power, with an efficiency of $66 \cdot 6$ per cent.

The output power is to be measured with an oscilloscope using a resistive dummy load, an RF ammeter, and a two-tone audio generator, whose tones are of equal amplitude and are not harmonically related.

The two-tone generator is applied to the transmitter and the carrier checked for full suppression. The input power is adjusted to give a mean RF output power under linear operation of 200 watts, measured into the resistive load—with the RF ammeter. At this condition, the peak to peak deflection of the output waveform is noted on the oscilloscope.

The tone is now replaced by speech from the microphone. The maximum deflection on the oscilloscope should not exceed that noted with the tone input. *See* Fig. 74a.

To see how the PEP of 400 watts is derived, let us work through an example.

150 watts DC input, at $66 \cdot 6$ per cent efficiency: —

$$\frac{150 \times 66 \cdot 6}{100} = 100 \text{ watts.}$$

If the dummy load R is $50 \, \Omega$ then the current required to produce 100 watts from:

$$I = \sqrt{\frac{P}{R}} \text{ is } I = \sqrt{\frac{100}{50}} = 1 \cdot 414 \text{ amps.}$$

The voltage across such a load under these conditions from

$$E = IR$$
$$E = 1 \cdot 414 \times 50$$
$$= 70 \cdot 7 \text{ volts.}$$

The peak voltage therefore at 100 per cent modulation is twice this value.

$$70 \cdot 7 \times 2 = 141 \cdot 4 \text{ peak volts,}$$

and the peak power is

$$P = \frac{E^2}{R}$$

$$P = \frac{141 \cdot 4 \times 141 \cdot 4}{50}$$

$$= 399 \cdot 9 \text{ i.e. } 400 \text{ watts PEP.}$$

The oscilloscope is a very useful and versatile instrument and there is no doubt that the presence of such an item makes servicing and testing much easier, since the waveform itself can be examined. It can be used to observe the depth of modulation in an AM transmitter by coupling a sample of the

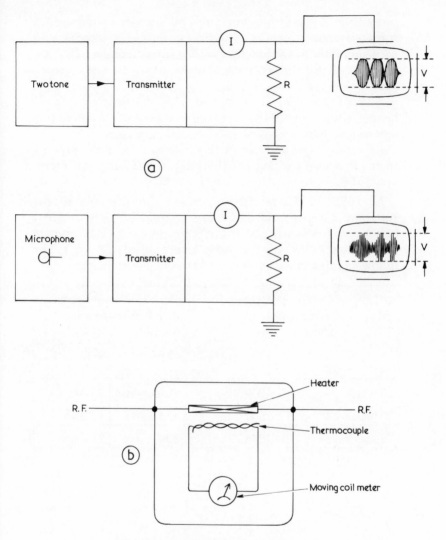

Fig. 74. a. Measuring peak envelope power
b. RF ammeter

modulated RF output to the vertical plates of the oscilloscope.
The resulting displayed waveform may then be monitored
continuously for correct envelope depth.

Finally we consider the RF ammeter used in the foregoing

171

procedure. Basically it is a moving coil instrument and as such will not respond to alternating RF currents. The movement is used therefore in conjunction with a thermocouple. This is a junction of two different kinds of metal which when heated produce a small direct current.

The RF signal to be measured is passed through a small heater in the instrument case and is therefore unaffected by any inductance. The heat produced by the current passes to the thermocouple and the small resulting current deflects the meter movement which is calibrated in milliamps or micro-amps of RF current (RMS). *See* Fig. 74b.

In 1982 the Home Office revised the schedule to the Amateur Licence and the more commonly used amateur bands now have their power limits expressed in dBW units of output power. (dBW is the power level relative to 1 watt). The following table shows the new and the old power listings: —

DC Input (old)	Carrier Power supplied to the antenna (new)	Peak Envelope Power supplied to the antenna for ssb operation (PEP)	
		old	new
10 watts	9 dBW	$26^2/_3$ watts	15 dBW
50 watts	16 dBW	133 watts	22 dBW
150 watts	10 dBW	400 watts	26 dBW

Chapter Eighteen

SAFETY

Safety is of paramount importance in the amateur's station, both to the operator and his visitors.

The best approach to safety, as with so many other facets of life, is common sense. Danger from shock increases with the current caused to flow through the body, so it is the capacity for delivery of current that determines whether a circuit is potentially harmful.

Low voltages are usually safe to the touch, a 12 volt supply is unlikely to cause any problems, whereas 240 volts is definitely unsafe to the touch. *Somewhere* between these two lies the boundary line between safety and danger. It is best to consider any electrical circuit which is directly or indirectly connected to the mains supply as a potential hazard.

Most of the equipment in an amateur station will be metal cased. This is good practice from the point of view of screening and the prevention of interference. However, metal screening can present a hazard in that being a conductor of electricity, if a fault develops that allows high voltage to be present on the case, that voltage will be conveyed to anyone touching it. For this reason a good earth is essential for all mains powered equipment, quite apart from its function as a signal earth.

Modern ring main installations will have approved earth returns connected. Older installations may have earths that have been taken to the cold water supply pipe. If this is the case, it is worth checking that a plastic pipe has not been inserted at some time between the rising main and the point at which the earth is connected.

If any doubt exists about the wiring, it should be checked by a qualified electrician. In any case, it is good practice to have the efficiency of the earth checked periodically.

The earth wire in a mains supply is connected via the earth pin in the plug directly to the chassis or case of the equipment.

Any current which appears on the case through a fault condition will be conducted by a low resistance path to earth. Any significant current flow in this way will cause the fuse to blow, thereby cutting off the source. This is a fail-safe situation. A blown fuse should always be regarded as a symptom of an electrical fault.

This is not always so, since fuses can fail through metal fatigue, but it is best to be cautious.

When changing fuses or investigating fuse failure—*switch off*. It is advisable to have a separate switch controlling the supply to all the station equipment so that it can be shut down with one movement. Such a switch should be of the double pole type, that is one which breaks both the mains supply lines. Other members of the household should be made aware of the location of the switch and its function.

Indicator lamps are a good safety feature too, since they act as a reminder that equipment is switched on. Lamps should be checked periodically and faulty ones replaced.

Careful attention should also be paid to insulation when constructing circuits that will be mains operated, such as power supplies for example. Insulated tools can be obtained, and these are very useful for the amateur who is going to undertake some construction work.

Large electrolytic capacitors can be a source of danger since, in good condition, they may hold a charge for several hours after the circuit of which they are part has been switched off.

It is wise to discharge any such capacitor by connecting the terminals to earth for a few seconds before carrying out any investigation. It is good practice to provide large capacitors with a permanent high resistance path to earth, through which they may discharge. This is usually done by means of a bleed resistor. The value is such that it does not prevent the normal function of the capacitor, but, when the supplies are switched off, allows it to discharge safely.

Headphones should always be removed when making adjustments, or investigating circuitry inside a receiver or transmitter which is connected to the mains, and meter test probes must be well insulated.

Aerials should not be connected directly to any mains or HT supply.

It does no harm to have a rubber mat to stand on when carrying out any such work.

Electricity need not be feared, but must be respected. Common sense is the best safeguard.

Chapter Nineteen

OPERATING PROCEDURES AND PRACTICES

A good deal of hard work goes into the process of obtaining a licence to operate an amateur radio station, and it must be born in mind that the licence itself is a privilege—not a right. Bad operating procedures, or contravention of the rules set out in the schedule attached to the licence, can result in the closing down of the station. This is by no means a common occurrence, since the vast majority of amateurs are highly responsible people, but it does happen, so it pays to study the schedule, and become familiar with the rules, which are sensible and very fair. The RSGB will provide, on request, full information on how to become a radio amateur, including a copy of the licence regulations. These are too complicated to be examined here, but there are two major points which should be mentioned.

The first is that the frequency bands allocated for amateur use are clearly defined, and operation outside those bands, whether or not it is accidental, is likely to result in the withdrawal of the licence. Out of band working will not be tolerated by the authorities.

The second vital item is the log book, which must be kept by every radio amateur, in the manner and type of book prescribed. This is not such a nuisance as some may think, since apart from satisfying the requirements, the log book is the amateur's diary, providing him with a complete record of his operations from the very first day. The log however is extremely important, and will from time to time be inspected by persons authorised by the Home Office. All contacts must be entered as they are made, with no gaps or erasures, and loose leaf books are not permitted. Suitable log books can be purchased or the operator may use a book ruled up by himself, provided of course that it conforms to the regulations. Vital information which *must* be logged is:—

1.   Date,
2.   Time, in G.M.T. start and finish of QSO (contact).

3. Frequency band used,
4. Mode of emission.
5. Call sign of station worked.

Other information may be recorded for the benefit of the operator as required. The manner in which call signs are allocated in various countries is complex and cannot be dealt with here, but the current state of affairs in the United Kingdom is relevant and fairly straightforward.

A successful RAE candidate receives the coveted written confirmation in due course, and the slip tells him the quality of his pass, in terms of distinctions and credits—though few will bother much, just as long as they have not failed. Many will use this to obtain a limited licence, the call sign for which will be G8 followed by three letters. G8 stations are strictly limited, in that they are not permitted to operate below 144MHz, so most operate on 2 metres, and possibly 70 cm.*

A full licence, enabling the licensee to operate on all amateur bands, can only be issued to those who have passed the Post Office Morse Test, set at twelve words per minute. This is well worth the effort, since the G4 licence permits working in CW or phone anywhere in the amateur bands, so putting the whole world at the fingertips of the operator. There are some G8 call signs which have only two letters after the figure, and these must not be confused with the G8 limited licence. A G8 two letter call sign is a very old one, and the licence is in fact the same as a G4. G2 and G3 calls will also be heard. These again are full licences which simply fit into the system before the current G4. There are also G6 calls, and again these are amateurs who have been licensed for many years.

The G in the call sign indicates that the station is operating from the UK, but a second letter is placed after the G if the location is other than in England. These letters are as follows:— GM—Scotland, GW—Wales, GU—Guernsey, GJ—Jersey, GI—Northern Ireland, GD—Isle of Man. These letters must be included in the call sign when a station is

* Probably during the currency of this book the G4 . . . and G8 . . . sequences will be used up. The Home Office will then issue new progressions.

operated in the country concerned, regardless of the home address of the licensee. Suffixes to the call sign are /M, /P, and /A. If a transmitter is used in a vehicle the /M is used, /P is for portable operation, and /A is used when operating the station from an address other than that shown on the licence.

A G5 call is a reciprocal licence, issued for a fixed period of time to a visitor to the UK who comes from a country which has a reciprocal arrangement, UK visitors to one of these countries can enjoy the same privilege. The question of how long an amateur should be satisfied with a limited licence is a vexed one, but there is no doubt about the fact that the operator who cannot work below 144MHz is missing out on a great deal of enjoyment and satisfaction.

Morse is a problem to many who attempt it, but in most cases this is due to sheer lack of self discipline. Occasional marathon sessions with practice key or recorded tape achieve very little, whereas two or three ten minute periods *every day* will soon bring results.

Another common mistake is in listening to very slow Morse, correctly spaced. Far better progress will be made by listening to recordings at twelve or fifteen words per minute, with a substantial gap between letters. The gap can be reduced gradually as time goes on, a fraction at a time. The writer (who works CW almost exclusively) has been most impressed by the Datong Morse Tutor, which generates random groups of five letters, numbers or a mixture of the two. This can be set up to send at any speed from six and a half to thirty seven words per minute, and the gap between the letters or numbers can be controlled anywhere between three seconds and the full calibrated speed. Used sensibly, this device is of tremendous value to the aspiring CW operator, but contrary to the opinion of some owners, the device cannot teach Morse—effort must be made. Simply switching a unit on and leaving it to run while the mind wanders, will not help. The spirit which exists among radio hams is a great thing, but nowhere is it stronger than among the CW fraternity. The advantages of transmission by Morse code are mainly that the bandwidth can be extremely narrow, also that tremendous distances can be covered on low power, and the fact that two operators who do not speak a word of each other's language can hold a conversation using the internationally recognised codes and abbreviations. An SSB transmission brought down to the bandwidth of CW would be quite unintelligible at the receiving end. Sections of each band are set apart for CW transmission, and in general this works very well indeed. Inconsiderate operators are sometimes heard working SSB in the CW sections, or vice versa, but such people are, thankfully, few.

Another aspect of amateur radio which is gaining popularity year by year is 'flea power', or QRP working. QRP operators work mainly in CW, and the expression QRP has been generally defined as working with a power output of less than five watts. Many operators today consider three watts to be the upper limit. One of the better known QRP rigs is the Heathkit HW8 (and its predecessor the HW7). This is sold by the Heath Company in kit form, and the kit, like all the others from Heath, is superb. It is a CW only transceiver, giving outputs of about two watts on 20 metres, one and a half on fifteen metres and so on.

Those who believe that linear amplifiers giving the full legal UK output of 400 watts PEP are necessary for efficient working, should try an HW8 with a good antenna and tuning unit — the experience could be most interesting. The quality of the antenna and the matching will have a far greater effect on the efficiency of a station than merely pouring on the power — which may well cause television or radio interference in the locality. QRP rigs can be run from a power supply for shack working, but another of their fascinating features is that they are very small, and can be operated from a battery. This makes them ideal for portable operation, since they can easily be packed into a suitcase and taken on holiday, or used on outings during good weather. A simple wire antenna, end-fed and matched by means of a small antenna tuning unit will work very well indeed.

CW is a fascinating part of amateur radio, and although it may seem difficult at first, the stage will eventually be reached where the letters themselves are heard, not the dots and dashes. Once this stage is reached, the true enjoyment of the mode is realised.

In both CW and SSB operation, the remnants of the once

179

mighty 'Q' Code are used. Some examples of this are included here, but the system has shrunk over the years to a mere shadow of its original form. It is used in most branches of radio communication, not just in the amateur bands.

QRL	—	I am busy.
QRL?	—	Are you busy? Used to check for a clear frequency before transmitting in CW.
QRM	—	I am experiencing interference.
QRN	—	I am bothered by static noise.
QRP	—	Now used to indicate low power. Original meaning was decreased power.
QRP?	—	Shall I decrease power?
QRQ	—	Send more quickly.
QRQ?	—	Shall I send more quickly?
QRS	—	The *cri de coeur* of the novice! Send slower.
QRS?	—	Shall I send slower?
QRT	—	Stop sending.
QRT?	—	Shall I stop sending? Now often used to indicate intended temporary close down of station.
QRV	—	I am ready.
QRV?	—	Are you ready?
QRZ?	—	Who is calling?
QSB	—	Your signal is fading.
QSL?	—	A request for acknowledgement. Generally a request for a QSL card to be sent.
QTH	—	My location is.

Note that the Q code consists of statements, which become questions if an interrogation mark follows them. Another common one is QSY, change frequency. This is usually sent as PSE QSY — please change frequency, when a station comes up on a frequency which is already in use. Since most amateurs spend as much time as short wave listeners (SWLs) prior to obtaining a licence they are usually conversant with the codes and abbreviations, which consequently present no problems. A study of books and RSGB publications will be very useful, however, since some operators are apt to employ bad procedures, which should be avoided at all costs. *A Guide to Amateur Radio*, published by the RSGB is an excellent example providing a sound basis upon which the operating procedures of a new station can safely be founded. Another extremely useful book from the same source is *The Morse Code for the Radio Amateur*.

A full discussion of operating procedures in all modes of emission would occupy a great deal more space than is available here, but some points regarding phone and CW should be mentioned. Good operating is largely a matter of common sense and good manners, and like a good driver, the first-class operator will inconvenience others as little as possible. Looking first at phone operation, one of the first points is that the recommended phonetics should be used wherever possible, because they are known to all operators and ambiguity in bad conditions is rendered unlikely. These phonetics listed below, are mentioned in the schedule to the licence as being 'recommended'. Others can be used, but they must not be objectionable or facetious. Since both words are open to wide interpretation, it is best to play safe.

Phonetic Alphabet.

A – ALPHA	B – BRAVO	C – CHARLIE
D – DELTA	E – ECHO	F – FOXTROT
G – GOLF	H – HOTEL	I – INDIA
J – JULIET	K – KILO	L – LIMA
M – MIKE	N – NOVEMBER	O – OSCAR
P – PAPPA	Q – QUEBEC	R – ROMEO
S – SIERRA	T – TANGO	U – UNIFORM
V – VICTOR	W – WHISKEY	X – X-RAY
Y – YANKEE	Z – ZULU	

The call signs G3UWJ and G4HWD would therefore be given respectively as Golf Three Uniform Whiskey Juliet, and Golf Four Hotel Whiskey Delta.

A radio contact (QSO) can be started either by listening for a station calling 'CQ', or by making a 'CQ' call. The CQ call ('seek you') is a way of calling for a contact, and can be qualified by calling CQ DX for a long distance contact, or CQ followed by the name of a country. The idea is to call CQ several times (not too many, about five will do) followed by the station call sign, repeated two or three times, and to repeat this with short listening breaks until contact is made—or enthusiasm wanes. If a contact is made, it is customary to thank the operator, and to exchange signal reports, names, and QTH (locations).

Signal reports in phone work consist of two numbers, the first will be from one to five, giving quality or readability, and the second, from one to nine, indicates signal strength as indicated by the 'S' meter on the receiver or transceiver.

It is always advisable to listen carefully before transmitting, to ensure that a frequency is not in use – and indeed to ask 'is this frequency in use please?' Contacts can be, and often are, confined to the details mentioned above, but some 'rag chews' can continue for hours!

In CW the CQ call will be much the same, but the signal report has three numbers, the last being indicative of tone quality. The readibility figure is of course the most important, and many amateurs pay little attention to 's' meter readings since meters vary in sensitivity, and such readings can be affected by various factors.

A typical 'rubber stamp' CW contact might run rather as follows:

DF6BFN DE G4HWD. GE OM ES TKS FER CALL! UR RST IS 589. HR QTH IS BATH NR BRISTOL. NAME HR IS GORDON. SO HW? DF6BFN DE G4HWD AR KN. which means, Good evening old man and thanks for call. Your signal report is 589 (figure group sent three times). My location is Bath near Bristol (name of town sent three times). Name here is Gordon (name sent three times). So how do you copy? AR means end of message. K means 'Transmit', and the N is added to let other stations know that a named station only should reply. The other station would reply giving the same information, and the QSO would then perhaps continue:

DF6BFN DE G4HWD. OK DR HANS ES MNI TKS FED RPRT. HR RIG IS YAESU FT901DM, ABT 70 WTTS TO ANT HQ1 WX HR IS SUNNY BUT COLD. WILL QSL VIA BURO SURE ES PSE UR QSL DR HANS? MNI TKS FER NICE QSO ES PSED TO MEET U HPE CUAGN SN. NW VY 73 ES GUD LUCK TO U ES FAMILY. 73 ES GB DR HANS DF 6BFN DE G4HWD AR VA KN, which is, O.K. dear Hans, and many thanks for fine business report, Here rig is Yaesu FT901DM, about 70 watts to HQI antenna. Weather is sunny but cold. Will QSL (send card) via bureau without fail, and please send your card dear Hans. Many thanks for the nice QSO and I am pleased to meet you. I hope to see you again soon. Now very best wishes and good luck to you and your family. Best wishes and goodbye dear Hans. AR — end of message, VA end of final transmission. RN — transmit. Abbreviations will also be used as much as possible in 'rag chews', but most are fairly obvious.

All this gives only a brief look at operating procedures, but the newcomer may be sure of a warm welcome and courteous treatment from virtually every station, so he will quickly feel at home. The very occasional contact with a rude or disagreeable ham (I have encountered only one) can be ignored.

Sadly there is room for no more, but we will end by wishing good luck, good DX and 73 to all hams, and all who are interested in the greatest of all hobbies.

RADIO SOCIETY OF GREAT BRITAIN PUBLICATIONS

A Guide to Amateur Radio (Pat Hawker, G3VA)
Amateur Radio Awards (ed. C. R. Emary, G5GH)
Amateur Radio Techniques (Pat Hawker, G3VA)
Amateur Radio Operating Manual (Ed. R. J. Eckersley, C4FTJ)
OSCAR—Amateur Radio Satellites (S. Caramanolis)
RSGB Amateur Radio Call Book
Radio Amateurs' Examination Manual (G. Benbow, G3HB)
Radio Communication Handbook, Vols 1 and 2
Radio Data Reference Book (T. G. Giles, G4CDY and G. R. Jessop, G6JP)
Test Equipment for the Radio Amateur (H. L. Gibson, G2BUP)
TVI Manual (B. Priestley)
VHF/UHF Manual (D. S. Evans, G3RPE and G R. Jessop, G6JP)

RSGB LOGBOOKS

Amateur Radio Logbook
Mobile Logbook
Receiving Station Logbook

RSGB WALL MAPS

Great Circle DX Map
IARU QTH Locator of Europe
QTH Locator of Western Europe
World Prefix Map

INDEX

A (ampere), 27 et seq.
Absorption wavemeter, 166, 167
Acceptor circuit, 57, 61
Adjacent channel interference, 111
Aerial (antenna), 104, 135 et seq.,
174, 179
ATU (tuning unit), 141, 143, 179
current and voltage distribution,
136, 137
directional beam, 138, 139
dummy load, 136, 168, 169
earthing system, 141
end effect, 136
end fed, 140–5
folded dipole, 139, 140
gain, 138, 139, 144
half-wave dipole, 136, 137
multiband, 141
parasitic elements, 139, 140
polarisation, 137, 152
quarter wave vertical, 141, 145
radiation resistance, 136
reflector and director, 139, 140
resonant length, 135, 136
trap, 142, 143
yagi, 138
AF (audio frequency), 20, 98 et seq.
AGC, amplified, 120
automatic gain control, 108, 157
delayed, 120
fast and slow, 120
Alternating current (AC), 45 et seq.
capacitance in, 51
inductance in, 51
peak value, 47
root mean square (RMS), 47
Amber, 21
Ammeter, 37
Ampere, 26, 27
Amplifier, AF, 85
class A, 86, 87

class B, 87
class C, 89
push-pull, 87, 88
RF, 85
transistor, 83
Amplitude modulation (AM), 98,
99, 100, 123
Analogue display, 118
Antenna, see Aerial, 135
Atom, 21, 68, 70
Avalanche effect, 92

Balun, 149, 150
Band, frequency, 20
valency, 23
Bandspread, 116, 117
Bandwidth, 60, 119
Barrier potential, 75, 79
Base, 79 et seq.
Battery, 24, 28, 43
Beam antenna, 138, 139
Beat note, 106, 118
Bel, unit of gain, 139
BFO (beat frequency oscillator),
106, 108, 118
Bias, 76 et seq.
potential divider, 81, 82
Books to read, 185
Breakdown voltage, 77
Buffer amplifier, 124–7
stage, 85

Calibration, 117
Call sign suffixes, 178
Call signs, 177
Capacitance (C), 39, 53
diode, 93
Capacitor, 40 et seq.
DC blocking, 82
electrolytic, 42, 43, 174

187

Capacitor —*cont.*
  ganged, 106, 107, 108
  trimmer, 95
  variable, 42, 103, 124
Capacitors, series and parallel, 56
Carbon atom, 21
Carrier frequency, 98, 99
  principle, 20
Cathode ray tube (CRT), 162
Cats whisker, 68, 103
Channels, frequency, 20
Charge, negative, 21, 23
  positive, 21, 23
Chirp, 126
CIO (carrier insertion oscillator), 108, 118
Clarifier, 117
Collector, 79 et seq.
Colour code, 29, 30
Communications receiver, 116, 117
Condenser, 44
Conduction band, 70, 71, 72
Conductor, 23 et seq., 70
Converter, 120
Copper, 23
Coulomb, 26, 27, 53
Coupled circuits, 63, 64
Co-valent bond, 68, 69, 70
Critical frequency, 155, 156
Crystal, 68, 69, 70
  calibrator, 117, 168
  set, 103
Current, 24–27
  conventional, 25
Cursor, 116, 117, 118
CW, 98, 106, 119, 123, 126
  learning, 178, 179, 181

dB (decibel), 139
DC power limits, 127
Demodulation, 112
Depletion layer, 75, 90
Detector, 104, 105, 106
  FM, 113
Deviation, 102, 115, 129
Dielectric constant — K, 157
Digital display, 118, 167
  frequency meter, 166

Diode, 77, 78, 79
  LED, 92, 93
  point contact, 103
  varactor, 92, 93
  zener, 92, 125
Dipole, half wave aerial, 136, 137
  radiation pattern, 137, 138, 140
Direct current — DC, 44 et seq.
Discriminator, FM, 114
Distortion, 86, 87, 89, 99
Donor atom, 72 et seq.
Doping, 72, 75
Drain, 90, 91
Drift, 119
Dummy load, 136, 168, 169
Dynamic resistance, 59, 60

Ear, 16
Earth, safety, 173, 174
Electrical charge, 21, 23, 24
Electrical theory, 10
Electricity, 19
Electromagnet, 35, 36
Electromagnetic radiation, 137, 151 et seq.
Electron, 21, 23, 24, 69 et seq.
EMF, back, 39, 52, 54
  electro-motive force, 28
Emission codes, 131, 132
Emitter, 79 et seq.
Energy band, 70, 71
  gap, 71, 76
Examination, Radio Amateur's (RAE), 9, 177

Fadeout, 157
Fading, 157
Farad (F), 40
Faraday, Michael, 37, 39, 40
Feeder (transmission line), 145 et seq.
  balanced and unbalanced, 145, 147
  characteristic impedance, 147
  coaxial, 145, 147
  mismatch, 146, 147
  reflected waves, 148
  standing waves, 148

Feeder — *cont.*
  twin, 145, 147
  velocity factor, 146
Filter, band pass, 63
  high pass, 62, 63
  low pass, 61, 63
Filters, 61, 62, 111
  crystal, 112
  mechanical, 112
Flux, magnetic, 46
Frequency, 14, 16–19
  band, 19, 20
  changer, 96, 97, 108
  counter, 166
  measurement, 162, 166
  modulation (FM), 102, 113, 128, 129
  shift, 160
FSK (frequency shift keying), 123
Fuse, 174

Gallium phosphide, 92
Gate, 90, 91
Generator, 45, 49
Germanium, 68, 70, 76
  diode, 103
Gramophone record, 18
Gravity, 25
Ground wave, 153, 154

Harmonic radiation, 166
Harmonics, 62, 86, 87, 118
Headphone, 107, 121
Heat, 19, 32, 33
Henry (H), 53
Hertz (Hz), 27, 46 et seq.
Heterodyne wavemeter, 168
Heterodyning, 96, 97, 131
Hole, 73 et seq.
Home Office, 9, 176
Hydrogen, 21

I (current), 27 et seq.
IF (intermediate frequency), 65, 108 et seq.
  gain, 119
Image interference, 109
Impedance (Z), 56, 59

Indicator lamps, 174
Indices, 47
Inductance, 37 et seq.
  self, 39, 53
Inductors, series and parallel, 55
Insulator, 20, 174
Interference, 10, 20, 109, 110, 111, 148, 158, 179
  transmitter, 9, 10, 132, 179
Intermediate frequency transformer, *see* IF
Ionised atom, 74
Ionosphere, 153–5
Ionospheric layers, 154–6
Ions, 75, 154

Lattice, 70, 73, 74
Leakage current, 76
LED (light emitting diode), 92, 119
Licence, A & B, 9
Licensing conditions, 9, 10
'Lift', 157
Light, 19
  waves, 151
Limiter, 114
Linear amplification, 86, 131
Log book, 176

Magnetic coupling, 65
  field, 35 et seq.
  flux, 46
Magnetism, 35, 36
Measurement, 10, 162
  current, 162
  frequency, 166
  power, 169
  resistance, 165
  voltage, 163
Meter, AC measurement, 165
  current, 162–5
  cut-out, 166
  FSD (full scale deflection), 163, 165
  moving coil, 162, 164, 168
  multiplier, 163, 164
  resistance, 165, 167
  RF ammeter, 169–72

Meter —*cont.*
  shunt, 163, 164
  voltage, 162–5
Microphone, 123, 128, 129
Mixer, 97
Modulation, 98, 99
  amplitude, 98, 99, 100
  envelope, 99
  frequency, 102
  high level, 128
  over, 99
Modulator, balanced, 129, 130
Morse code, 98, 106, 123
  learning, 178
  test, 9, 10, 177
Morse key, 98, 124, 126, 127
Morse, Samuel, 98
MUF, 155
Multimeter, 165
Multiplier, frequency, 96, 124, 127, 129
Muting, 122

NBFM, 102, 129
Netting, 124
Neutron, 21
Noise limiter, 120, 121
Non-linear amplification, 87, 89
N-type semiconductor, 72 et seq.
Nucleus, 21, 23, 69

Ohm, 26, 27
Ohm's law, 28, 34, 35
  formulae derivations, 34
Operating procedures and practices, 10, 176
Operating, starting a contact (QSO), 181
  typical CW contact, 182, 183
  use of phonetic alphabet, 181
  use of 'Q' code, 180
OSCAR satellite, 160
Oscillator, 60, 94
  Colpitts, 94, 96
  frequency changing, 97, 108
  local, 108
  RF, 94, 124

Quartz controlled, 95, 96, 117, 125, 131
VFO (variable frequency), 94, 124, 129, 131
Oscilloscope, 162, 169

PA (power amplifier), 124–8
Parallel tuned circuit, 59, 60, 104, 128
Pentavalent impurity, 72, 73
Phase difference, 49 et seq.
Phonetic alphabet, 181
Pinch-off voltage, 90
PN junction, 74
Post office, 10
Potential difference (PD), 24
Potentiometer, 30, 31
  variable, 30, 31, 106
Power (P), 32, 33, 66
  DC input, 169
Power gain, transistor, 83, 84
Power, PEP (peak envelope power), 169, 170, 171
Product detector, 112, 113
Propagation, 135, 151
  and Antennas, 10
  Summary, 158, 159
Proton, 21
P-type semiconductor, 72 et seq.
Pump, 24, 25
Pythagoras' theorem, 50

'Q' code, 180
Q (coulomb), 26
Q (of tuned circuit), 60, 104
QRP (low power), 179
Quartz crystal, 95

RADAR, 123
Radio Amateur's Examination (RAE), 9
Radio communication, 20
  control, 123
  receiver, 10, 103 et seq.
Radio Society of Great Britain, 11, 176, 180, 185
Radio waves, 20, 151
Reactance (X), 53, 62

Reciprocal call sign, 178
Recorder output, 121
Rectifier, 78
    RF, 104, 105, 106
Reflection, sky wave, 153-7
Reflectometer, 148
Refraction, 157
Rejector circuit, 59, 61
Repeaters, 160, 161
Resistance (R), 26 et seq.
    colour code, 29, 30
Resistor, 28, 29
    bleed, 174
Resistors in series and parallel, 31,
    32, 33
Resonance, 56, 57, 59
    in aerials, 135

RF (radio frequency), 20, 98 et seq.
RFC (radio frequency choke), 113
RF gain, 119
RIT (receiver independent tuning)
    control, 133
RTTY (radio teletype), 123

Safety, 173
Satellites, 160
Screening, 132
Second channel interference, 110
Selectivity, 104, 111, 131
    variable, 119
Semiconductor, 10, 24, 68 et seq.
Series tuned circuit, 56, 60
Shell, 23
Signal strength meter, 119
Silicon, 68, 70, 76
Sine wave, 18, 45 et seq.
Single sideband (SSB), 101, 112,
    119, 123
SI units, 26
Skip distance, 155
Sky wave, 153, 154, 155
Slope detection, 114
Solar cell, 160
Solenoid, 36
Sound waves, 13, 16, 19
Source (FET), 90, 91
Square wave, 45, 86

SSB filter method, 129
Stability, frequency, 119, 124
Standing wave ratio meter, 148
S units, 120
Sunspot cycle, 155
Superhet, 108, 109
Switch, double pole, 174
SWR (standing wave ratio), 148, 149

Telegraphy, 98
Telephone, 13, 19
    tin-can, 13, 14, 19
Temperature - effect on frequency,
    124
Thermal runaway, 84
Thermocouple, 171, 172
Tone burst, 160
Transceiver, 133
Transducer, 19
Transformer, 65
Transistor, 68 et seq.
    bipolar, 90
    common base, 80, 81
    common collector, 85
    common emitter, 83
    FET, 90
    heat sink, 84
    npn, 79 et seq.
    pnp, 79 et seq.
    power, 84
    unipolar, 90
Transmission line, see Feeder, 145
    et seq.
Transmitter, 10, 62, 123 et seq.,
    152 et seq.
    basic AM, 124
    output impedance, 128
    SSB, 129, 130
    VHF, 129, 130
Transponder, 160
Trap, aerial, 142, 143
TRF receiver, 105, 106, 107
Trivalent impurity, 73
Troposphere, 153, 157
Tuning fork, 13, 14, 16, 17, 18
TV receiver, 62

Valency band, 23, 70, 71

Valve, 68, 92, 105, 128
Varactor (varicap), 92, 129
Vector diagram, 50, 52, 59
Velocity factor, 146
   of radio waves, 151
VHF bands, 128, 129
Video transmission, 123
Violin, 17, 18
Voice, human, 19
Volt, 26, 27, 28
Voltage stabilisation, 92
VOX (voice operated transmitter),
   134

Water analogy, 24–27
Watt (W), 34
Wavelength, 18, 152, 153

X-rays, 151

Yagi, 138

Z (impedance), 56, 59
Zener diode, 92, 125